VISCERAL MUSCLE
Its Structure and Function

TERTIARY LEVEL BIOLOGY

A series covering selected areas of biology at advanced
undergraduate level. While designed specifically for course
options at this level within Universities and
Polytechnics, the series will be of great value to
specialists and research workers in other fields who require
a knowledge of the essentials of a subject.

Other titles in the series:

TERTIARY LEVEL BIOLOGY

Visceral Muscle

Its Structure and Function

HENRY HUDDART, Ph.D.

Lecturer in Biological Sciences,
University of Lancaster

and

STEPHEN HUNT, Ph.D.

Lecturer in Biological Sciences,
University of Lancaster

A HALSTED PRESS BOOK

John Wiley and Sons,

New York

Blackie & Son Limited
Bishopbriggs
Glasgow G64 2NZ

450 Edgware Road
London W2 1EG

Published in the U.S.A. by
Halsted Press,
a Division of John Wiley and Sons Inc.,
New York

Library of Congress Cataloging in Publication Data
Huddart, Henry.
Visceral muscle, its structure and function.

(Tertiary level biology)
Bibliography: p.
Includes index.
1. Viscera—Muscle. I. Hunt, Stephen, joint
author. II. Title. DNLM: 1. Muscle, Smooth.
WE500 B883v.
QP321.H813   591.1'852   75-42464
ISBN 0-470-15221-4

Printed in Great Britain by
Thomson Litho Ltd., East Kilbride, Scotland

Contents

INTRODUCTION

This text covers the main aspects of activation and contraction of visceral muscles, and the modulation of these activities by extrinsic and intrinsic control mechanisms. The text is aimed at second and third-year undergraduates in physiology, pre-clinical medicine and zoology, and it may form an introduction to postgraduate studies. Wherever possible, a structure/function approach has been adopted, since anatomy and physiology are so interrelated that to ignore one of these aspects is to present an imperfect picture. More advanced postgraduate studies are amply served in advanced works published recently (*Handbook of Physiology*, 1968; Bulbring *et al.*, 1970).

The literature on visceral muscle is vast, confusing and often contradictory, and it grows increasingly each year. Much of the confusion in visceral-muscle physiology is due to the inherent variability of the tissue itself. Whereas skeletal (or somatic) muscle is designed to execute a particular task, such as shortening to move an appendage or to exert tension at a point, visceral muscles are designed to execute a wide variety of quite different tasks. Since visceral-muscle cells constitute the walls or operative parts of the major visceral organs, their physiology is inextricably linked to the operation of the individual organ which they constitute. This has resulted in an immense diversity of contractile behaviour. The kind of mechanical activity required in a pulsatile artery, or in the bladder, uterus, ureter, seminal vesicles, pilomotor system, the gut, lung and chromatophores is very diverse, the activity of any one muscle being optimum for the efficient operation of that particular organ. Whereas spontaneous rhythmic contractions may be ideal for an artery or for the gut, this activity would be of little use in the bladder, uterus or seminal vesicles. As a result of the diversity of organ systems in animals, we find that the activity of visceral muscles is as varied as are the activities of the organs they constitute. This is a major reason for visceral muscles being so much more variable in their electrical and mechanical behaviour than is skeletal (or somatic) muscle.

Since visceral muscles are so intimately involved in visceral functions, it is hardly surprising that their activity varies enormously from time to time, since they are so greatly affected by the general physiological state of the animal. Most visceral muscles exhibit both short and long-term cyclical variations in their activity, and this is particularly obvious in muscles in organs affected by hormonal changes. In addition, the visceral muscles of the gut and vascular system undergo diurnal changes in their activity, related to cyclical variation in blood noradrenaline levels. This natural cyclical variation in activity is another major factor contributing to the confusion and contradictions in the research literature.

A further point of confusion surrounds exactly what is meant by 'smooth muscle'. On a strict structural classification, smooth muscle simply refers to those muscles which are clearly not striated, including the visceral muscles of vertebrates but excluding the visceral muscles of many invertebrates. Visceral muscle is a more sensible, functionally-orientated term, since it includes all the unstriated muscles of the vertebrate viscera, and also the striated visceral muscles of the arthropods. Since the approach of this text is functional, the collective term 'visceral muscle' will be used as appropriate.

The great confusion surrounding the variability of visceral muscles was somewhat clarified when Bozler (1948) proposed that these muscles, no matter how diverse, could be considered as belonging to one of two major types. *Multi-unit muscles* are non-rhythmically active, being galvanized into contraction by impulses in their extrinsic nerve supply from the central nervous system. Characteristic of this type of muscle are many blood vessels, the muscular coats of much of the male genital tract, such as the vas deferens, the seminal vesicles, the urinary tract, such as the ureters and bladder, and the insect and molluscan rectum. *Single-unit muscles* are in general rhythmically and spontaneously active, being driven by spontaneous rhythmic depolarizations of the neurones in their intrinsic plexuses. Here the extrinsic nerve supply simply acts to modulate (but *not* to initiate) the level of activity of the intrinsic neurones. Characteristic of this division are the muscles of much of the alimentary canal and some pulsatile blood vessels. These muscle types thus differ in the site of origin of their activity: intrinsic in single-unit muscles and extrinsic in multi-unit muscles. It is now clear that this simple classification really defines only the extremes of physiological variation. Some intermediates are known, but the Bozler classification has proved to be of immense use and will be retained with some caution in our text.

Both authors are most grateful to their colleagues and research associates, who have generously loaned some of their unpublished work and electron micrographs. In particular, we wish to thank Dr. A. J. Syson of Lanchester Polytechnic for much new and exciting data on excitation-contraction coupling in mammalian muscle, and Mr. K. Oates of the University of Lancaster for his advice and criticism on electron microscopy and for his interest in this topic.

HENRY HUDDART
STEPHEN HUNT
Department of Biological Sciences
University of Lancaster

CHAPTER 1

THE FINE STRUCTURE
OF VISCERAL MUSCLE

Visceral muscle is a tissue, and hence by definition is composed of cells, all of which are rather similar in organization to one another. Because the function of this tissue is a specialized one, namely mechanical activity, the cells too are highly specialized. All muscle cells, be they from skeletal muscle, cardiac muscle or visceral muscle, have a basically similar organization, when considered in the most simple manner. That is to say, they have that feature which is common to all cells: a cytoplasm contained by a limiting outer membrane, within which is an elaborate system of fibrous proteins whose function is to bring about contraction at the cellular and hence tissue levels. Contractile activity is mediated metabolically by a supporting cast of cellular organelles, similar to those found in most metabolically active cells. While all muscle cells have this common pattern, it is the details of the way in which the fibrous elements are arranged within the cytoplasm of the cell, and their relationship with the other cell organelles, that make the striking differences between visceral muscles and the skeletal and related types. This chapter is intended to give an account of the structure of visceral and other non-striated or smooth-muscle cells, as seen through the electron microscope. While much of what we say will refer to vertebrate visceral-muscle cells, it must be remembered that outside the arthropods, where the majority of muscle cells are of the striated type similar to those seen in vertebrate skeletal muscle, most invertebrates have a large proportion of their body musculature composed of tissues which in many ways resemble vertebrate visceral muscle; these, however, are more suitably called 'smooth' muscles, because they do not occur viscerally and are non-striated. We shall not therefore feel inhibited from drawing examples from smooth and visceral muscles in invertebrate phyla. We shall consider the ultrastructure of the innervation of the muscle tissue in Chapter 3. Readers who are unfamiliar with general cellular ultrastructure are recommended to refer to an introductory work, such as that by Toner and Carr (1971).

1

Tissue and cell

Smooth-muscle tissue in the vertebrates is found principally in visceral and vascular situations. The majority of the hollow organs in the vertebrate body have an outer longitudinal coat of smooth muscle and an inner circular coat. In blood vessels, on the other hand, the muscles are most usually arranged in a spiral fashion, wound around the vessel like the coils of a spring. The muscles thus make an angle with the long axis of the vessel, and this angle varies in indirect proportion with the diameter of the vessel. Therefore as the diameter decreases, as it does in the transition from arteries to arterioles, the angle increases, with the result that in some very tiny arterioles the muscle is arranged around the vessel in what is to all intents and purposes a circular manner. A more detailed account of the disposition of visceral muscle in the vertebrate body is given in Chapter 2.

Within the muscle coats, the muscle cells are arranged in bundles of quite irregular shape, sometimes branching and embedded in a sheath of connective tissue (figure 6.1). These bundles vary in diameter but in most cases would probably approximate to 100 μm.

Most basic texts describe visceral-muscle cells as being spindle-shaped with a single centrally-placed nucleus. In fact this is far from true. Although there is usually only one nucleus, its position within the cell, and from cell to cell in a particular tissue, can vary quite widely; this is particularly noticeable in many invertebrate smooth muscles. Serial sectioning of smooth-muscle cell bundles for electron microscopy has shown that the cells have very irregular shapes and sizes. Although idealized as fusiform in outline, in reality they appear with highly irregular longitudinal contours, often with quite blunted ends and with cross-sections varying from polyhedra through circles, triangles, squares, ellipses to flattened ribbons. Some cells, such as those found in arteries or in the intestinal tract walls of certain molluscs also seem to branch. Some caution, however, is necessary, since the failure to fix muscle under relaxing conditions may cause the appearance of great irregularity.

Published values of cell dimensions vary quite widely, with the added complication that the dimensions of cells in the contracted state are, not unnaturally, somewhat different from those of cells which are relaxed. In general, cell lengths fall somewhere between about 30 μm (arteriole) and 450 μm (vas deferens), while diameters lie between 2 and 6 μm.

Cell-cell relationships

The macroscopic phenomenon of muscle contraction is only possible because the individual muscle cells are tied together as an effector unit in the bundles in a specific relationship to each other.

In muscles such as vas deferens, the cells do taper classically at their ends. Here they pack together by overlapping, with their thinner ends against the thicker nuclear central regions of neighbouring cells; but it is common to find end-to-end arrangements as well as interdigitated systems. The numerical relationships between cells in the bundles vary, but electron micrographic serial sectioning suggests that over the length of a single muscle cell, in tissues such as vas deferens or taenia coli, each cell will have about 12 surrounding cells. However, because of overlap, at any particular position along the cell this number will usually be reduced to about 6. As the cells tend to interweave, the picture is difficult to obtain accurately, even with serial sectioning.

Individual cells in a bundle are usually separated by a gap of about 60 nm which narrows considerably in the region of intercellular junctions. This space between the cells is filled with an extracellular matrix of acid mucopolysaccharides, glycoproteins and collagen, and with elastin fibres

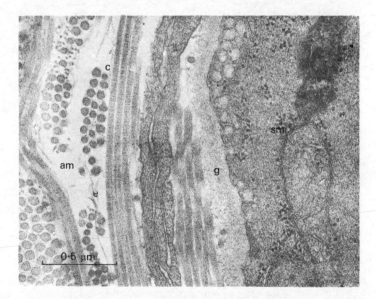

Figure 1.1 Electron micrograph of rat ileal smooth muscle to show the extracellular space. Note the bundles of collagen fibres (c) in transverse and longitudinal section, running at right angles to each other through an almost clear matrix of acid mucopolysaccharide (am) which contains a few elastic filaments (e). A hazy glycocalyx of glycoprotein and acid mucopoly-saccharide (g) extends for some distance into the extracellular matrix from the surface of the smooth muscle cell (sm). The repeat periodicity on the collagen fibres is of the order of 64 nm.

with entrapped quantities of water, salts and other low-molecular weight metabolites. Between the muscle cell bundles, the extracellular space opens out and the matrix becomes more extensive. Here collagen fibres are often clearly in evidence (figure 1.1), and single axons or small axon bundles may sometimes be seen.

Intercellular junctions

As mentioned above, the major part of the surfaces of most of the cells in a smooth-muscle cell bundle is separated by a gap tens of nanometres wide, filled with so-called *basement membrane material.* There are, however, regions in which adjacent muscle cells come into close contact with each other, forming contacts which take a variety of forms. These

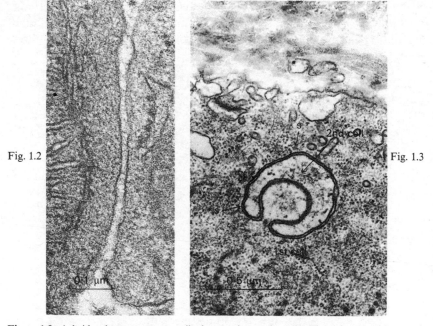

Fig. 1.2   Fig. 1.3

Figure 1.2. A bridge between two rat ileal smooth-muscle cells. There is no evidence of structural modification of the opposed cell membranes, but there is apparently some fine fibrillar material in the space between the cells.

Figure 1.3. An intrusion of one smooth-muscle cell into another in pharyngeal retractor muscle of *Buccinum undatum.* The intrusion is in transverse section, and mates closely with the contact cell. The membranes apparently make bridge contacts at several points, and some junctions of the septate type (s) are apparent.

contacts can have the function of providing electro-chemical continuity between the cells in a bundle or, in some cases, may simply provide a mechanical connection between cells. In general, there are three main types of contact, for which the literature presents a somewhat confused nomenclature. Thus we find one type of contact called a *bridge*, in which the separation between cells decreases to about 10–20 nm, the cell membranes forming an area of close approach of about 0·3 µm (figure 1.2). Included in this type of contact are the so-called *intrusions*, in which an out-pushing of one cell inserts into an invagination of an adjacent cell, rather in the manner of a ball-and-socket joint. Figure 1.3 shows an intrusion of this type found in a molluscan smooth muscle. Bridges and intrusions together probably account for 5–10% of the total cell-membrane surface area in a smooth-muscle cell bundle.

Of greater physiological importance are the cell contacts, variously called *nexuses, close junctions, tight junctions* or *gap junctions*. These structures appear to be rather sensitive to the method of fixation used for electron microscopy, and may separate if conditions are not exactly right; these conditions vary from tissue to tissue. In the region of a nexus, the gap between the cells disappears and the outer layer of the two adjacent cell membranes may even fuse. What we see in the electron microscope,

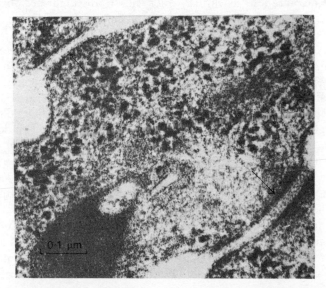

Figure 1.4. Two nexus junctions between a pair of rat ileal smooth-muscle cells. Plate courtesy of Dr. A. J. Syson.

then, is essentially a pentalaminar membrane system, rather than two trilaminar membranes with a light gap separating them (figure 1.4). The pentalaminar system is usually about 14 nm wide, and normally there is no unusual modification of the cytoplasmic faces of the cell membranes. Nexuses are normally of the order of 0·1–0·2 μm across, and serial sectioning of cells suggests that they may occupy up to 6% of the surface area of the smooth-muscle cell. We have already said that nexuses are thought to be the morphological loci of electrical coupling; but it must be added that visceral muscles are known (e.g. canine duodenum longitudinal muscle or vas deferens) in which there is ample physiological evidence for electrical coupling, and yet nexuses are absent or rare. This does not mean that the nexus is not a site of electrical coupling, but it may mean that other less easily recognized loci exist as well. Nexuses may occur either where cells simply lie closely adjacent to each other or in the cellular intrusions described above (Gabella, 1973).

At the fine structural level, the most readily apparent and distinctive forms of intercellular junction are the 'attachment plaques', which are probably synonomous with the desmosomes commonly found in other types of tissue (Gabella, 1973). These junctions are characterized by dense formations of fibrous material on the cytoplasmic faces of the two opposed membranes—a feature which clearly distinguishes them from the nexus type of junction. Where an attachment plaque forms, the gap between the adjacent cells can be as much as 30 nm. In this gap there is usually a layer of electron-dense material lying in the extracellular matrix (figure 1.5). It is a frequent feature of attachment plaques in visceral and other smooth-muscle cells that part of the contractile protein system of the cell, the thin filaments, can be demonstrated to form bundles in the region of the attachment plaques, and to diverge towards and merge with them. Attachment plaques of the type we have described are symmetrical, in the sense that the cytoplasmic thickening of one cell has a corresponding thickening in the cell to which it is attached. However, non-symmetrical attachment plaques are also of common occurrence which, morphologically at least, resemble in every aspect one half of the symmetrical plaque, with a patch of dense material on the cytoplasmic side of the cell membrane. Linked with these non-symmetrical plaques are bundles of myofilaments and, when contracted cells are fixed for electron microscopy, the plaques are seen to have been pulled into the cell, carrying with them an infolding of the cell membrane (figure 1.6).

It seems likely that the attachment plaques fulfil the dual function of

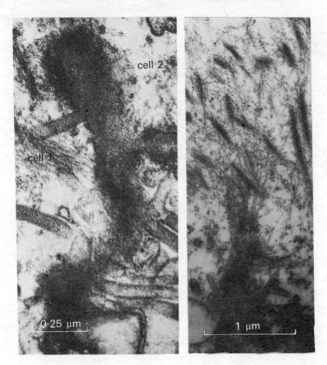

Fig. 1.5

Figure 1.5. A symmetrical attachment plaque in visceral smooth muscle from the intestine of the mollusc *Buccinum undatum*. Dense material totally obscures the membranes and the extracellular space in the immediate region of the plaque.

Figure 1.6. A single attachment plaque from the same source as figure 1.5. The plaque is pulled into the body of the cell, apparently by the thin filaments which deviate into it.

maintaining the general structural integrity of the muscle tissue and providing in part the anchorage points for the contractile protein system, so that movement of the cell as a whole, and not just within the cytoplasm, can be achieved.

The cell membrane

As we have seen, the surface of the smooth-muscle cell is frequently involved in permanent or semi-permanent interactions with other adjacent cells. However, those regions of the cell membrane not involved in this latter type of activity by no means always present to the electron microscopist the simple profile of uninterrupted stretches of simple

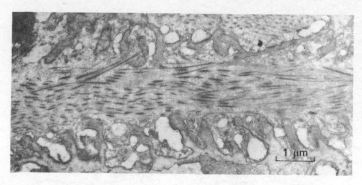

Figure 1.7. Longitudinal section of a molluscan-gut smooth-muscle cell. Note the highly irregular outline with numerous large invaginations, as well as the smaller finger-like inpushings of the cell membrane into the cytoplasm of the cell.

trilaminar membrane (figure 1.7). Rather the membrane often follows an irregular contour, with quite frequent invaginations—both large and small—into the cytoplasm of the cell.

Large irregular sometimes-branching invaginations of the plasmalemma are often found. These are seen in a highly elaborate form in the gut muscle of certain molluscs (figure 1.8), where we have found them apparently penetrated by the extracellular matrix material which comprises the 'basement membrane' around the cells. However, collagen fibres do not penetrate far into these invaginations. At their innermost points of penetration into the body of the cell, they seem to contain a fuzzy electron-dense material, resembling the material of the non-symmetrical attachment plaques with which they are sometimes associated. Slightly less elaborate invaginations are also found in vertebrate visceral muscles, for example in ano-coccygeus (Gabella, 1973), where the absence of collagen fibres in the extracellular material filling them is very noticeable (figure 1.9). Of course, these large invaginations cause a very significant increase in the surface area of the cell and, incidentally, also increase the volume of the extracellular space—two factors which are likely to have important physiological consequences. Whether these invaginations are analogous to the T-system of skeletal muscle, and whether they take any part in the inward transmission of excitation are still matters of dispute.

One of the most characteristic features of vertebrate visceral-muscle membranes are the flask-shaped *caveolae intracellulares* (micropinocytotic vesicles) (figure 1.10). These invaginations of the cell membrane

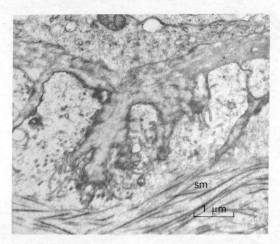

Figure 1.8. Detailed view of a molluscan smooth-muscle cell (sm), showing a large branching invagination which carries numerous tubular, finger-like inpushings of the cell membrane.

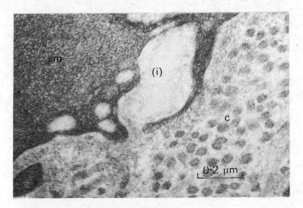

Figure 1.9. Cell membrane invaginations (i) into a rat ileal smooth-muscle cell (sm). The extracellular space included by this invagination contains no collagen fibres such as are seen (c) in the general extracellular space,

are surprisingly uniform in size and, in electron micrographs, tend to give the impression of running in rows at quite regular intervals. Most workers seem to agree that the *caveolae* are in direct contact with the outside of the cell, i.e. with the extracellular matrix. It is fairly generally believed that they represent a region where ionic exchange between the extracellular space and the cytoplasm of the cell is particularly facilitated. The mouth of the *caveola* is about 20 nm across, where it narrows to

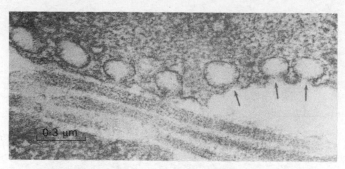

Figure 1.10. *Caveolae* in vertebrate smooth-muscle cell membranes (arrowed).

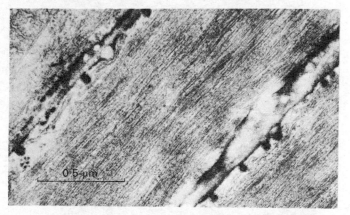

Figure 1.11. *Caveolae* in guinea-pig ileal circular muscle after treatment with lanthanum hydroxide. Lanthanum fills the *caveolae* but does not enter the cytoplasm of the cell. Courtesy of Dr. G. Gabella.

the neck of the 'flask', and the *caveolae* of a smooth-muscle cell may increase its effective surface area by more than 70%—as, for example, in taenia coli muscle. The collagen fibrils of the basement membrane do not penetrate the *caveolae* (Gabella, 1973).

Muscle which is loaded with uranium salts prior to fixation and sectioning for electron microscopy shows the heavy metal bound all over the surface of the cells and completely penetrating the *caveolae*, while not entering the cytoplasm; when colloidal lanthanum is used, the entire extracellular matrix and caveolar spaces are filled, and the cytoplasm is again not penetrated (figure 1.11). If washing-out of lanthanum is allowed to take place, it is nevertheless usually retained quite strongly as a thin layer along the plasma membrane, while completely filling the *caveolae*, which suggests an extracellular composition within the *caveolae* rather different from that in the general extracellular space (Gabella, 1973).

It has been suggested that *caveolae* are perhaps related in function to the transverse tubular systems of skeletal and cardiac muscles. Certainly striated muscle cells in lizard ventricular tissue or rat heart atrium have been recognized in which transverse tubules are absent but numerous *caveolae* are present. The argument that transverse tubules are unnecessary in visceral muscle because the low degree of penetration of the *caveolae* is adequate to the requirements of the small visceral-muscle cell requires a cautious approach.

Caveolae as such seem not to be found in the smooth-muscle cells of invertebrates. We have, however, found in molluscan smooth muscle, particularly in that of the gut but also elsewhere, an extensive development of finger-like invaginations of the cell membrane (figure 1.8), many of which penetrate relatively deeply into the cytoplasm of the cell. The structures have a very constant diameter, about 75 nm, and enter the cell either directly from the plasma membrane or quite frequently from the large irregular invaginations already described (figure 1.8). Unlike *caveolae*, these invaginations seem to contain material somewhat more densely staining and granular than that of the extracellular matrix and resembling that of the cell membrane glyocalyx or surface polysaccharide protein coat, which in these cells is thicker and denser than in vertebrate muscle. Many of these long finger-like invaginations run straight into the cytoplasm, but many others seem to turn a sharp angle just below the cell membrane, and then to run parallel to the surface of the cell. The result of this is that we commonly see many of these tubules in transverse section, lying in clusters near to the periphery of the cell (figure 1.8) and running among sub-surface cisternae. In longitudinal section we have frequently observed a spiral pattern on the outside (extracellular) walls of the invaginations (figure 1.12) which, on closer examination, appears to originate in spirally arranged rows of discrete particles bound to the membrane surface. Such systems have been seen in other situations and termed *coated* or *striated vesicles*; they occur, for example, in the visceral muscle of turtle oviduct, vena cava and aorta (Somlyo *et al.*, 1971).

Some authors distinguish the *caveolae* from micropinocytotic (plasmalemmal) vesicles, while others are less definite about this (see the reviews of Gabella (1973) and Burnstock (1970) for a comparison of attitudes). The major grounds which do seem to justify this distinction are the smaller size and homogeneity of the *caveolae*, and the frequent presence of a diaphragm across the mouth of micropinocytotic vesicles.

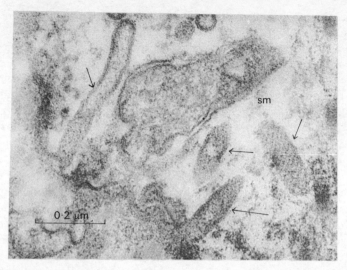

Figure 1.12. Detailed view of 'coated' finger-like invaginations (arrowed) of the cell membrane of molluscan intestinal wall muscle (sm). Note the spiral arrangement of particles on the extracellular face of the membranes. Compare these structures with the *caveolae* of vertebrate visceral muscle.

Sub-membrane organization

We have already dealt in one respect with the region immediately inside the visceral-muscle cell membrane, in the sense that the *caveolae* clearly enter this region. Morphologically speaking, however, they are but extensions of the extracellular space. In close association with the cell membrane and the *caveolae*, although not in direct connection with them, is usually an elaborate system of membranous sacs, tubes, cisternae and vesicles, mostly of quite irregular size, shape and distribution. Much controversy surrounds this system of vesicles, since several workers now hold it to be an endoplasmic reticulum functionally related to the sarcoplasmic reticulum lateral-tubule system of vertebrate and arthropod striated muscle. Figure 1.13 shows this so-called 'sarcoplasmic reticulum' in vertebrate smooth muscle, while figure 1.14 shows similar reticulum in a molluscan smooth muscle. There is often an apparent tendency for the sacs and cisternae to be particularly developed along the long axis of the cell in the superficial regions, and they often seem to form between rows of *caveolae*, as a gutter lying under rows of *caveolae*, or in sheets immediately beneath the plasma membrane. Although this reticulum is most frequently composed of smooth membranes, areas of the ribosomal

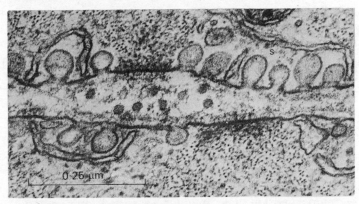

Figure 1.13. 'Sarcoplasmic reticulum' (s) in vertebrate visceral muscle. The reticulum consists of sacs and tubules in the region of the cell membrane and the *caveolae*. Courtesy of Dr. G. Gabella.

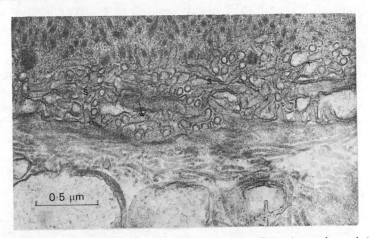

Figure 1.14. 'Sarcoplasmic reticulum' (s) in an invertebrate (molluscan) smooth muscle (compare with figure 1.13). Here the membraneous sacs are intimately associated with the tubular invaginations (arrowed) shown in figure 1.8.

rough endoplasmic reticulum type do occur in the superficial regions of the cell; however, the more typical endoplasmic reticular manifestations of labyrinth smooth and rough membranes and of Golgi bodies usually occur most prominently in the cytoplasmic regions around the two contractile element-free areas which are distributed symmetrically at the ends of the long axis of the nucleus (the *nuclear poles*).

The relationship of the sub-surface system of cisternae, sacs and tubules to the cell membrane, and in particular to the *caveolae*, has led to speculation that this sarcoplasmic reticulum-like system fulfils some function in the import, storage and transport of calcium in visceral muscle. Direct fine-structural evidence for this tends to be lacking, although some biochemical evidence is available. We shall discuss this problem in its physiological context in greater detail in Chapter 5.

Mitochondria

The mitochondria of visceral-muscle cells are typical of mitochondria from many other non-muscle cell situations in their general morphology, although their cristae are frequently more abundantly folded than in the average mitochondrion. This might be expected for a situation where transduction of chemical energy to mechanical energy is the prime function of the whole cell. Spherical and irregularly shaped mitochondria are not uncommonly seen, but the majority seem to be prolate ellipsoids about $0.5\,\mu$m long and $0.2\,\mu$m wide. Longer filiform types are sometimes found lying between the myofilament bundles of the cell.

Distribution of mitochondria in smooth-muscle cells seems to vary quite dramatically from tissue to tissue. In some tissues, such as guinea pig ileum, there are numerous peripherally-placed mitochondria closely associated with the *caveolae* and the sub-surface cisternae. By contrast, in a tissue where we might expect a similar situation, e.g. rat ileum, the

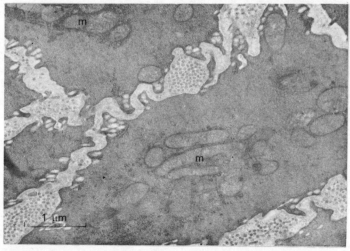

Figure 1.15. Central groups of mitochondria (m) in cells of rat ileal smooth muscle.

mitochondria are centrally placed in bundles in the cell (figure 1.15). This is also true of molluscan pharyngeal retractor muscle (figure 1.16), where closely packed bundles of up to 15 mitochondria have been observed deep in the cytoplasm of the cell among the myofilaments. Mitochondria are frequently found in large numbers orientated along the long axis of the smooth-muscle cell, at either end of the nucleus.

It is worth remembering that in striated muscles the mitochondria are much more precisely distributed, being located mostly near the Z-bands in the general region of the triad system where T-tubules and lateral sarcoplasmic reticulum come into contact. In this respect, there may be parallels in smooth muscle where *caveolae*, mitochondria and sub-surface cisternae interact.

Nuclei

Visceral-muscle nuclei are frequently described as cigar-shaped and centrally placed in the cell. In fact, this is by no means always the case. They are often quite irregular in outline, particularly when seen in transverse section (figure 6.1), and may be eccentrically distributed in the cytoplasm, both laterally and longitudinally (figure 1.17).

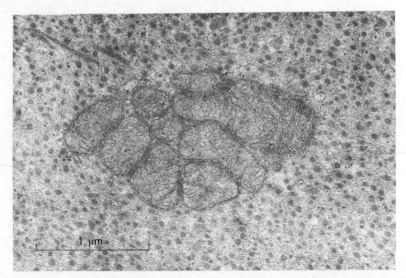

Figure 1.16. Situation comparable to that of figure 1.15 in molluscan pharyngeal retractor muscle. The mitochondria are densely packed and almost entirely exclude the cytoplasm of the cell.

The smooth muscle of rat ileum presents the classical appearance, with a centrally placed nucleus, ellipsoidal in shape, with the characteristic indentations which seem to spiral around it, and which intensify on contraction and disappear on extension of the cell (figure 6.1). The nucleoplasm of smooth-muscle nuclei is characteristically rather pale in the electron microscope, with darker-staining peripheral chromatin and large prominent nucleoli, of which there may be more than one (figure 1.17). Surrounding the nucleus is a double trilaminar membrane system, collectively termed the *nuclear membrane*, which delimits, between the paired membranes, a perinuclear cisterna about 13 nm across. A common feature of the outer surface of the nuclear membrane, i.e. that in contact with the cytoplasm, are bound ribosomes (figure 1.17). The nuclear membrane is usually penetrated at intervals by pores.

The contractile protein system

A cell is an integrated unit in which all parts and organelles have a functional role to play. It is therefore clearly illogical to refer to any structure within a cell as being the most important. Even so, in any fine-

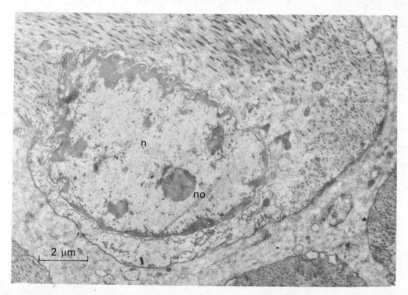

Figure 1.17. A large excentrically placed nucleus (n) in molluscan opercular somatic smooth muscle. The nucleus has a highly convoluted outline, a large nucleolus (no) and lies in an area largely free of myofilaments. Compare this figure with figure 6.1.

structural account of muscle cells, that part of the cell which brings about the mechanical contraction of the cell as the end product of its metabolic activity, must receive very important treatment; it is to be expected that this structure, the myofilamentary apparatus, will figure prominently in electron micrographs of smooth-muscle cells.

The contractile filament system of skeletal and cardiac muscle is composed of a highly ordered arrangement of regularly overlapping thick filaments (myosin) and thin filaments (actin) (see figure 7.2). We discuss the biochemical and biophysical aspects of the organization of the proteins of the thick and thin filaments in some detail in Chapter 7, where we note that the proteins of the contractile apparatus of non-striated muscle are essentially similar to those of skeletal and cardiac muscle, but that the thick filaments of the latter muscles are organized from the protein myosin in a rather different manner from the myosin filaments of smooth muscle.

As we mention in Chapter 7, the question of the ultrastructure of the contractile protein system in visceral muscle has, until recently, been surrounded with uncertainty, argument and counter-argument. The cause of this controversy was the fundamental point of whether *in vivo* the myosin molecules in visceral muscle existed as dispersed elements or as thick filaments comparable with the thick filaments of striated muscles. This is, of course, a matter of great importance since the mechanism of contraction, which has been worked out in some detail for skeletal muscles, rather depends upon the existence of well-organized thick filaments of a relatively stable and permanent character (Chapter 8). The majority of workers, until very recently, were unable to demonstrate consistently and conclusively in the electron microscope the existence of thick filaments in the visceral muscle of vertebrates, in spite of the detection by chemical methods of the existence of large amounts of myosin in the tissues, and the physiological evidence for a sliding filament mechanism. It now appears that this failure resided in the lack of an adequate technique for the preparation of visceral muscle for electron microscopy. Thick filaments may now be demonstrated in tissue sections in a quite routine manner (Somlyo *et al.*, 1973). Some authors still argue that if such care has to be taken, then the thick-filament situation cannot be the same as it is in skeletal muscle, and that the thick filaments may still have to be considered in visceral muscle as transient structures. Indeed the thick filaments are organized in a rather different manner (Chapter 7) but it is likely that the basic principles are very similar to those obtaining in skeletal muscle. The smooth muscles of invertebrates

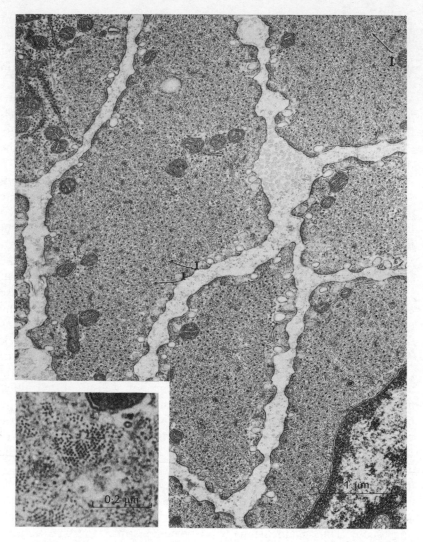

Figure 1.18. Transverse section of smooth-muscle cells in rabbit portal-anterior mesenteric vein. The thick filaments are regularly spaced and are surrounded by large numbers of thin filaments. Intermediate filaments (I, arrows) and associated dense bodies are visible. Courtesy of Dr. A. P. Somlyo.

Figure 1.19 (inset). Transverse section of mouse intestinal smooth muscle, showing a detailed view of a group of thin filaments arranged in a hexagonal lattice. Courtesy of Dr. H-G. Heuman.

usually present no problems where demonstration of thick filaments is concerned. These differences in demonstrability of thick filaments seem to reside partly in differences in proteins, which function to dictate the structural mode of assembly and to stabilize the myosin polymers (Chapter 7).

Thin filaments

Once the problem of demonstrating the thick filaments has been overcome, visceral and vascular smooth muscle in transverse section, for most tissues of vertebrate origin, appears as in figure 1.18. Fairly regularly spaced thick filaments are surrounded by large numbers of thin filaments, which appear at first sight to be quite randomly distributed. This photograph should be compared with one of a skeletal muscle cell (figure 6.4B) in which, by contrast, the very regular arrangement of thick and thin filaments should be noted.

The thin filaments in the smooth-muscle cell are composed of actin (Chapter 7) and closely resemble those of skeletal muscle. The thin filaments range in diameter from 4·8 to 8·0 nm, and seem to have a ratio to thick filaments of the order of 15:1. These are values for rabbit portal-anterior mesenteric vein, but are rather typical of most vertebrate material. In the invertebrates, the thin filaments of smooth muscle are essentially similar (figures 1.16 and 6.4A). For example, in gastropod hypobranchial gland muscle the thin filaments have diameters in the range 9·0–11·0 nm, and have a ratio to thick filaments of 30:1. Although the initial impression given by the thin filaments is that of a random distribution, they tend actually to arrange themselves in complete or incomplete circular orbits around the thick filaments, as well as in isolated rosette clusters and in single or double rows running between rows of thick filaments (figures 1.16 and 6.4A). Centre-to-centre spacings for thick to thin filaments are of the order of 25·7 nm in mammalian smooth muscle, but these distances are difficult to estimate in invertebrate materials because of the very variable diameter of the thick filaments. The ordering of actin thin filaments into bundles with regular lattice-like arrangements is particularly obvious in relaxed muscle fibres (Heuman, 1973), with relatively extended areas in which there is compact packing in hexagonal array (figure 1.19). In these areas, filaments have centre-to-centre spacings of 4–5 nm. In *Mytilus* (Mussel) anterior byssus retractor muscle, the packing is hexagonal in multi-rows with two-dimensional branching. Areas free from thin filaments are a characteristic feature of transverse sections of smooth muscle (figures 1.18 and 6.4A). In contracted muscle, the situation we have just described seems to change, with thick and thin filaments far more randonly distributed relative to themselves and to each other.

In longitudinal section, the thin filaments are seen to be arranged predominantly along the long axis of the cell and running parallel to the thick filaments (figure 1.20). This is, however, only the generalized trend, and the course followed by the thin filaments may be quite sinusoidal, with sudden wide swings away from the general line of the cell towards an attachment plaque or a dense body (figure 1.6). The filaments appear very thin and tenuous in longitudinal section, and it is difficult to see much detail of their substructure. However, some sections of material of invertebrate origin (figure 1.21) show what can be construed either as a very short repeat of about 5 nm with a slight obliquity, or else as a beaded organization, which is believed to be the case for actin filaments prepared *in vitro* (Chapter 7). In transverse section, actin filaments of vertebrate visceral muscle tend to appear as of even electron density, but those of invertebrate smooth muscle may frequently appear to have an

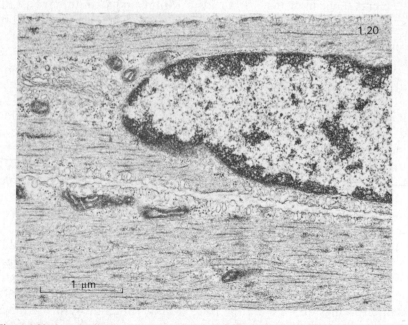

Figure 1.20. Longitudinal section of a smooth-muscle cell from rabbit portal-anterior mesenteric vein, showing thick and thin filaments. Courtesy of Dr. A. P. Somlyo.

electron-lucent core (figure 1.22)—a feature shared by actin filaments in some vertebrate and arthropod skeletal muscles.

Thick filaments

As we have already mentioned, the problems associated with the demonstration of myosin thick filaments in electron micrographs of vertebrate visceral muscle have proved to be a great hindrance to the understanding of the way in which the contractile apparatus is organized and functions. To achieve this demonstration, several factors require careful attention: water swelling and penetration of chloride ions must be reduced to a minimum during preparation of the tissue for fixation; osmium tetroxide alone cannot be used, but may be combined with glutaraldehyde or formol; temperature is probably also important (Somlyo *et al.*, 1973; Heuman, 1973; Shoenberg, 1973).

Under these optimum conditions, thick filaments of between 13 and 18 nm diameter can be seen readily in thin sections of vertebrate smooth muscle (figure 1.18). The profiles of these thick filaments are rather irregular, and views under higher magnification suggest that they are composed of sub-units (Chapter 7). It is extremely difficult to orientate a small piece of embedded tissue sufficiently accurately for exact

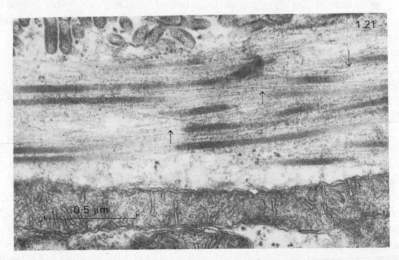

Figure 1.21. Longitudinal section of molluscan gut muscle. Note the beaded appearance of the thin filaments (arrowed).

longitudinal section, in such a way that entire thick filaments can be included in a single section; however, the occasional combination of good luck and good management (which makes research a feasible enterprise) has led to a general consensus of opinion that thick filaments in visceral and vascular smooth-muscle cells are probably about 13–15 μm long. Unlike thick filaments in invertebrate smooth muscles, those of vertebrate material show little or no evidence of regularly repeating longitudinal organization. While the ends of many longitudinally sectioned thick filaments are seen to be tapered, it is difficult to say whether this is a reality of structure or an artefact of sectioning. As in the case of the thin filaments, the predominant orientation in longitudinal section is of all thick filaments parallel. However, divergences from this pattern are quite common, and some workers have reported apparent splitting of filaments and formation of V-junctions; again these may be sectioning artefacts.

Thick filaments in invertebrate smooth muscles have a more cylindrical cross-section (figure 1.23) and wider ranges of diameters than those in vertebrate tissues (figures 6.4A and 1.18). For example, the thick filaments in the smooth muscle of holothurian tube feet range from 25 to 60 nm in diameter when measured in transverse section, those of gastropod hypobranchial gland muscle from 15 to 120 nm, and those of *Mytilus* (Mussel) byssus retractor muscle from 10 to 75 nm. A detailed study of the latter muscle has shown that the range of diameters recorded is due to sectioning at different points of filaments of approximately constant length, which taper away uniformly on either side of a central zone of constant diameter. Thick filaments in byssus retractor muscle are about 25 μm long (Sobieszek, 1973). Thick filaments in invertebrate smooth muscles and some pseudostriated muscles usually show a longitudinal periodicity (figure 1.24) of between 12 and 13 nm, attributable partly to the presence of the protein paramyosin or tropomyosin A (Chapter 7).

Although perhaps not immediately obvious, the thick filaments of both vertebrate and invertebrate visceral, vascular and somatic smooth muscle are not distributed completely randomly (cf. figures 1.18 and 6.4A). In rabbit portal-anterior mesenteric vein, the thick filaments are usually separated from each other by approximately 60 × 80 nm, in a quasi-rectangular lattice. In anterior byssus retractor muscle, there is short-range hexagonal packing of thick filaments. In *Buccinum* pharyngeal retractor (in transverse section) the thick filaments tend to run in curving rows (in transverse section).

Cross-bridges between thick and thin filaments

In Chapters 7 and 8 we discuss aggregation of the myosin elements to produce thick filaments, which occurs in such a manner that the globular regions of the myosin molecules project from the surfaces of the filaments as lateral projections. We also discuss the manner in which these myosin 'heads' function as cross-bridges with actin filaments, mediate the actual sliding of the thick and thin filaments relative to each other, and bring about contraction.

In striated muscle, the regularity of alignment of the thick and thin filaments, and the stability of the fibrillar system to glycerine treatment (which removes obscuring background protein) allow us to visualize the cross-bridges between the filaments relatively easily. Because of the irregularity of organization in smooth muscle, the greater spacing of the filaments, and the poor preservation of visceral muscle in glycerine (invertebrate muscle is less sensitive), cross-bridges are identified much less easily. Even so, it is still possible to detect on the thick filaments of vertebrate smooth muscle, lateral projections suggestive of those seen on the myosin filaments of striated muscle (figure 1.25). Similar cross-bridges can be seen in invertebrate muscle (figure 1.26), in both longitudinal and transverse section.

Intermediate filaments

Aside from the thin actin and thick myosin filaments of vertebrate smooth muscle, a third class of filaments can usually be observed, with diameters of about 10 nm, intermediate between the thin and the thick (figure 1.18). These are described as having regular round cross-sections, sometimes with an electron-lucent core. They tend to occur in bundles, to run longitudinally through the cell for distances of several micrometers, and are better preserved during osmium fixation alone than are the thin filaments. In this respect, they seem better able to survive the slings and arrows of outrageous fixation than either the thin or the thick filaments, and can be seen to persist after treatment which has to a large extent caused loss of ultrastructurally recognizable polymeric actin or myosin. We comment on the probable protein composition of these filaments in Chapter 7. Transverse sections of vertebrate smooth-muscle cells show a tendency for these filaments to congregate around the periphery of the dense bodies (see below) and, in fact, they can be extracted from muscle cells apparently attached to dense bodies. Possibly they are part of the

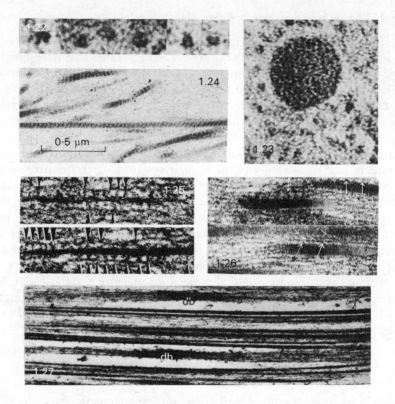

Figure 1.22. Thin filaments in transverse section of molluscan hypobranchial gland smooth muscle, showing the irregular outline and electron-lucent core.

Figure 1.23. A thick filament from molluscan smooth muscle in transverse section. Note the regular outline and appearance of a sub-structure of small units as well as the orbit of thin filaments.

Figure 1.24. Typical invertebrate smooth-muscle thick filaments in longitudinal section, showing the regular 12–13 nm periodicity.

Figure 1.25. Longitudinal section of thick filaments in rabbit portal-anterior mesenteric vein smooth muscle, showing thick filaments with cross-bridges to adjacent thick filaments. Courtesy of Dr. A. P. Somlyo.

Figure 1.26. A region similar to that shown in figure 1,25, but in a molluscan muscle. The cross-bridges (arrowed) are less distinct here.

Figure 1.27. Dense bodies (db) in mussel (*Mytilus*) anterior byssus retractor muscle, showing the bundles of thin filaments emerging on either side of the dense body. Courtesy of Dr. A. Sobieszek.

cytoskeleton of the cell. Some large areas in the central regions of the cell are occupied by intermediate filaments to the almost total exclusion of actin and myosin filaments (Somlyo *et al.*, 1973).

Intermediate filaments in invertebrate smooth muscle have attracted the comments of few workers, but the higher upper limits which have been quoted for the diameters of actin thin filaments in these tissues may require reappraisal in this respect. However, the impression gained from examining transverse-section micrographs of molluscan muscles, such as figure 6.4A, is that the distribution of myofilament diameters is bi- rather than trimodal. Moreover, micrographs of glycerinated anterior byssus retractor muscle published by Sobieszek (1973) show no sign of filaments of diameters intermediate between the paramyosin-myosin thick filaments and the actin thin filaments.

Dense bodies

Under the heading of 'intercellular junctions', we have already mentioned the presence of attachment plaques. These exist as masses of dense amorphous or slightly fibrous material on the inner surfaces of regions of smooth-muscle cell membranes, at points of close contact between adjacent cells. We also said that these desmosome-like bodies could also exist assymetrically on membranes, in the absence of a contact with an adjacent cell. We further mentioned that thin filaments enter these attachment plaques. In fact, it is the thin actin filaments which are involved and, apart from these membrane-associated bodies, very similar spindle-shaped structures, about $0 \cdot 2$–$0 \cdot 4$ μm long and $0 \cdot 01$ μm wide in vertebrates and up to 2 μm long in invertebrates, are found in free form, i.e. not attached to membranes in the cytoplasm of the smooth-muscle cells associated with the thin filaments (figure 1.18). These free units are called *dense bodies*.

Dense bodies are found in both vertebrate and invertebrate smooth muscle, as well as in the obliquely striated body-wall muscles of *Ascaris*. They vary in number from tissue to tissue from being, for example, quite prominent in the musculature of the intestinal wall of the whelk *Buccinum* to being only rarely seen in the pharynx retractor muscle of the same species (figures 1.8 and 1.16); similarly, in the case of the rat, they are more numerous in the aortic wall than in the uterus or the bladder.

The present view of the role of dense bodies is that they function analogously to the Z-bands of striated muscles, i.e. essentially as an anchorage point by which organized and orientated contractility can be maintained at the cellular level, while the membrane-bound attachment plaques serve to integrate contractile activity throughout the tissue. We have already referred to the manner in which the membrane becomes pulled inwards by the filaments and their associated plaques during contraction (figure 1.8).

Force has been added to the argument that the dense bodies are functionally related to the Z-discs of striated muscle by recent detailed observations of anterior byssus retractor muscle in *Mytilus*. Here careful fixation and sectioning has defined sarcomeres, 50 μm long at non-overlap, in which there are three to four 25 μm-long thick filaments and about seventy 11 μm-long thin filaments attached to either half of a dense body (figure 1.27) (Sobieszek, 1973). Evidence is also available to suggest that, like the actin filaments on either side of a skeletal-muscle Z-disc, the thin filaments on either side of a dense body are oppositely polarized. Some workers dispute the existence of dense bodies, suggesting that they are in reality artefacts of the fixation process, resulting from disrupted thick filaments (see Chapter 8).

The dual role of visceral-muscle cells

Visceral-muscle tissue always appears to be lacking in sufficient fibroblast cells to account for what is often a very extensive extra-cellular matrix and basement membrane development around the muscle cells. It has been postulated that at early stages of development, even before the myofilamentary system has been elaborated, visceral-muscle cells are highly active in the biosynthesis of extracellular materials, including collagen, mucopolysaccharides and possibly elastin. Some evidence to support this is now available from tissue culture work. Developing tissues such as postnatal aorta show extreme synthetic activity on the part of the vascular smooth-muscle cells in respect of extra-cellular protein production. This type of activity is not shared by skeletal or cardiac-muscle cells.

Striated muscles and smooth muscles represent two evolutionary extremes of the possible ways of organizing a contractile apparatus based on myosin and actin within a cell. Between these two extremes lie, uneasily, a number of muscles which in structure seem to be neither truly striated nor truly non-striated. As an example, we may take the oblique or pseudostriated muscles of cephalopods, with their regular rhomboidal sarcomeres, T-tubules, lateral endoplasmic reticulum and precisely aligned bands of dense bodies which are not joined but, en masse, resemble much more closely a Z-disc than the scattered dense bodies of other molluscan smooth muscles. Slow-acting visceral muscle of insect gut might be expected to be similar to muscle in other phyla, but it is in fact striated. The structure, however, is diversified possibly in relation to its situation and function. The Z-discs are irregular in outline, in some species almost breaking up into separate dense body-like units; the sarcoplasmic reticulum is reduced to small flattened sacules, the T-tubules are short, and the thick filaments are surrounded by rosettes of twelve thin filaments rather than by the more common six. Could this last point represent a step of convergent evolution in the direction of the multiple thin filament arrays of smooth muscles?

CHAPTER 2

THE INNERVATION
OF VISCERAL MUSCLE

The problem of exactly how visceral muscle is activated involves several complex factors:

(1) The almost universal presence of an extrinsic innervation from axons whose pathways lie in the autonomic division of the central nervous system.

(2) The presence of intrinsic nerve elements in a variety of visceral plexuses buried within the smooth-muscle tissue.

(3) The presence, within the plexuses, of afferent sensory neurones responsible for the mediation of inhibitory reflexes.

These afferent neurones are routed out from the effector organs through the intrinsic ganglia and innervating nerves into the central nervous system.

As was mentioned in the introduction, visceral muscles can be roughly divided into multi-unit types or single-unit types; these differ fundamentally in the way in which the central nervous system exercises control over their activities. In multi-unit (non-rhythmic) muscles, contraction results from impulses arising in or routed through the extrinsic nerve supply. In terms of neural control, these muscles are the nearest to skeletal muscle in its voluntary control by the somatic central nervous system. In general, single-unit muscles are automatically rhythmically active. The excitation leading to their contraction arises in the tissue, in the intrinsic nerve cells of the ganglia within the various plexuses. With this type of organization, the extrinsic nerve supply from the autonomic nervous system acts simply to modulate the inherent automatic discharge rhythmicity of the plexus neurones by direct preganglionic excitatory or inhibitory synaptic action. The method of neural control here is thus analogous with the extrinsic nervous control of neurogenic hearts. Some care must, however, be exercised in applying in these cases a term so traditional to cardiac muscle. In both types of visceral muscle, the central nervous system is in ultimate control, but the pathways are more complex and less direct in single-unit muscles. In some cases, visceral muscles may be truly myogenic, the excitation leading to contraction arising within the muscle fibre membrane. These continue to show spontaneous activity in the presence of drugs which abolish neural activity. In these muscles, central-nervous-system modulation of contraction takes place by a direct action on the muscle fibre membrane. This organization is similar to that in an innervated myogenic heart.

Some visceral muscles do not fit easily into such a convenient classification. The visceral muscle of the cockroach rectum is such a case. This muscle is innervated by the proctodeal branches of the cercal nerve from the sixth abdominal ganglion (Brown and Nagai, 1969). Impulses in this nerve set up postsynaptic excitatory potentials in the muscle fibres, leading to contraction—a situation similar to that in normal skeletal muscle. However, some muscle fibres of this preparation are spontaneously active, even when the preparation is denervated. The spontaneous activity in this case is intrinsic, almost certainly generated within the enteric nerve plexus. This preparation thus exhibits activity characteristic of both single-unit and multi-unit muscle, since the whole organ system possesses two separate populations of fibres with different activation control.

The autonomic nervous system

As was mentioned above, the various organs of the body which are composed of visceral muscle receive an innervation from the autonomic nervous system. The autonomic nervous system can be defined generally as being composed of all the various scattered efferent pathways (i.e. pathways leaving the central nervous system) which have a synapse in a ganglion or ganglion-like structure *outside* the physical confines of the central nervous system. Defined in this way, the autonomic system includes the efferent innervation of the main viscera, such as the gut, the vascular system, the urino-genital system, the lungs, the liver, the kidneys, the iris and ciliary body, and many glandular structures such as skin sweat glands, salivary and lachrymal glands, and endocrine structures such as the adrenals and pancreas. The autonomic system also innervates the heart. In addition to the efferent supply to the above structures, afferent axons from peripherally placed sensory neurones also run into the central nervous system, routed through the autonomic nerves, such as those to the gut, bladder and genital system.

The autonomic nervous system comprises essentially a two-neurone pathway from the central nervous system to the innervated organ. One neurone is situated within the central nervous system, its axon leaving the neuraxis (the spinal cord or cranial nerves) and synapsing on a second neurone in a peripherally-placed ganglion. The second neurone has its axon running from the ganglion to the visceral muscle or gland. The terms *preganglionic* and *postganglionic* have been coined for these neurones, the former describing the neurone of central origin, the latter describing the neurone arising within the peripheral ganglion. In general, preganglionic fibres are of large diameter and myelinated, while postganglionic fibres are very thin and unmyelinated. There are, however, exceptions to this generalization.

It became clear towards the end of the last century that the autonomic nervous system innervating the gut consisted of two quite separate divisions. These divisions differed not only in their anatomical routing from the central nervous system, but also in the type of chemical transmitter agent liberated by their neurones and to which their neurones were sensitive. The early anatomical terms for these autonomic divisions were (1) the *craniosacral* system and (2) the *thoracolumbar* system. The craniosacral system consisted of axons leaving the cranial nerves to cephalic structures, such as lachrymal and salivary glands and the eye, and also the vagal supply to the heart, lungs, and anterior part of the gut, including the liver, pancreas and kidneys. The rest of this system consisted of axons leaving the sacral segments of the spinal cord, routed through the pelvic nerves, innervating the posterior part of the gut, the bladder and the genital system. This craniosacral system, which is now known to release the chemical transmitter acetylcholine, was found to be largely excitatory on smooth muscle. The more modern name for this collection of neurones is the *parasympathetic nervous system*. The thoracolumbar system consisted of neurones routed out from the central nervous system in the thoracic and lumbar sections of the spinal cord, innervating the gut, the heart and vascular system, and various glands via the paravertebral ganglia lying on either side of the spinal cord. This system, releasing mainly noradrenaline (norepinephrine), is largely inhibitory on smooth muscle. The modern name for this collection of neurones is the *sympathetic nervous system*. In general, activity in the sympathetic nervous system is associated with increased alertness and activity, while activity in the parasympathetic nervous system is associated with the more vegetative states of sleep and digestion, where bodily activities tend to be rather depressed. A very readable introductory account of the autonomic nervous system can be found in Wyburn (1960).

Visceral-muscle innervation in the vertebrates

The distribution of the different parasympathetic and sympathetic innervations of visceral muscle, and the nature of the responses they induce in the animal kingdom are far too varied and still too much a matter of dispute for any attempt at a comprehensive review here. The vertebrates, which form a relatively cohesive group, still show great variation in the autonomic control of the viscera, and in the responses of different visceral muscles to a particular innervation. In general terms, the sympathetic nervous system is excitatory on vascular smooth muscle but inhibitory on gut musculature, while the parasympathetic system may be either excitatory or inhibitory, depending upon the organ involved. Generally, the parasympathetic system is excitatory on gut muscle but inhibitory on the vascular system. Throughout the vertebrates it is difficult to see any consistent evolutionary pattern relating a particular peripheral structure and its response to a particular innervation. Table 2.1 attempts to present a rough picture of the main features of parasympathetic and sympathetic control of visceral muscle in the vertebrates. The great variability of nervous and pharmacological actions is apparent in these few examples. The action of the vagus nerve in the vertebrates is a classic case of variable action. In cyclostome fishes, vagal action either relaxes or has no action on the gut; hence the vagal endings cannot be cholinergic, since acetylcholine itself is excitatory on the gut of these fishes. Here it is possible that some sympathetic postganglionic fibres have joined the vagus on their route to the gut, and it may be these which have been stimulated. In elasmobranchs, vagal stimulation may be either excitatory or inhibitory, which is also the case in the teleosts. Again in the amphibia, vagal stimulation may be excitatory or inhibitory, the vagus apparently carrying both types of fibre to the gut, as would appear to be the case also in birds and mammals. Much of the confusion surrounding the effects of autonomic nerve stimulation is due to the fact that these nerves are often mixed. In many cases (see later) it is clear that sympathetic postganglionic fibres join with parasympathetic preganglionic fibres in the last part of their travel to various parts of the gut. Since this is so, then whether reports indicate that sympathetic noradrenaline-liberating fibres or parasympathetic acetylcholine-liberating fibres or both are stimulated is entirely dependent upon circumstantial conditions.

Table 2.1.—The innervation and some of the pharmacological properties of some selected visceral muscles.

Animal and muscle	Parasympathetic innervation	Sympathetic innervation	Acetylcholine	Noradrenaline	Reference
Toad stomach muscle	Vagus inhibitory, transmitter unknown	Cervical excitatory splanchnic inhibitory	+ (excitatory)	+ (inhibitory)	Campbell (1969)
Cat and rabbit bladder, prostate, vas deferens and seminal vesicle	Sacral supply via pelvic nerve—inhibitory	Hypogastric nerve—excitatory		+ (excitatory)	Sjostrand (1965)
Sleepy lizard bladder	Sacral plexus both excitatory and inhibitory	Not established	+ (excitatory)	+ (inhibitory)	Burnstock and Wood (1967) McLean and Burnstock (1967)
Cat portal vein	Vagal effects are weak and inconsistent	Splanchnic nerves excitatory	+ (slightly excitatory)	+ (excitatory)	Johansson and Ljung (1967)
Teleost stomach (cod)	Vagus both adrenergic excitatory and cholinergic excitatory	Splanchnic nerves adrenergic and excitatory	+ (excitatory)	+ (excitatory)	Nilsson and Fange (1969)
Chick and pigeon gizzard	Vagus excitatory	Perivascular nerve both inhibitory and excitatory			Bennett (1970)
Carp alimentary canal	Vagus both excitatory and inhibitory	Sympathetic system inhibitory	+ (excitatory)	+ (inhibitory)	Ito and Kuriyama (1971b)
Trout intestine	Vagus excitatory and inhibitory	Splanchnic nerve excitatory (cholinergic)	+ (excitatory)	+ (inhibitory)	Ito and Kuriyama (1971b) Campbell and Burnstock (1968)
Chick mesenteric artery	Excitatory	Inhibitory	+ (excitatory)	+ (inhibitory)	Bolton (1968)

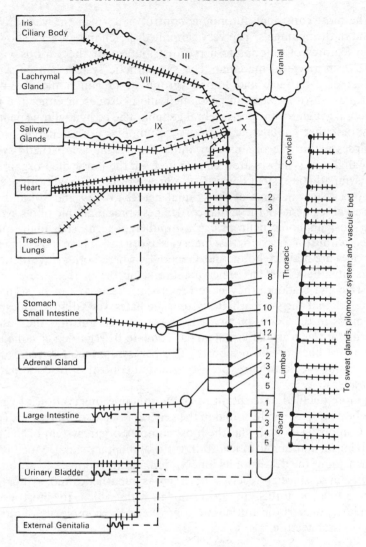

Figure 2.1. Rough sketch showing the distribution of mammalian autonomic nerves. Sympathetic preganglionic neurones (————) leave the spinal cord in the thoracic and lumbar spinal nerves. They enter the paravertebral (O——O) or prevertebral (●) ganglia, where they connect to postganglionic neurones (++++) which innervate the viscera. Parasympathetic preganglionic neurones (– – – –) leave the spinal cord in the sacral and cranial nerves, making contact with postganglionic neurones (~ ~ ~O) either in the tissue itself (☐———) or in remote ganglia (~~~|———). Redrawn and simplified from Campbell (1970).

The most consistent autonomic control pattern in the vertebrates is found in the mammals. The very simplified sketch in figure 2.1 shows the main features of autonomic nerve distribution to the various visceral muscles in a typical mammal. The sympathetic system can be seen to consist of a thoracic and lumbar group of preganglionic cholinergic neurones which synapse on postganglionic neurones in either the paravertebral ganglia or the prevertebral ganglia. The postganglionic neurones may be either cholinergic or, more commonly, noradrenergic at their synapses within the organs they innervate. In the sympathetic system, the ganglionic synapses are well remote from the innervated organs. The parasympathetic system consists of a group of fibres leaving the central nervous system in some of the cranial nerves (3, 7, 9 and 10) and from the sacral segments of the spinal cord. These preganglionic fibres synapse onto postganglionic neurones in a number of terminal ganglia placed either within or near to the innervated organs. These postganglionic neurones are mainly cholinergic. Generally, where both sympathetic and parasympathetic systems innervate the same organ (e.g. the stomach, intestine and bladder), they have reciprocal effects. It can be seen from figure 2.1 that the sympathetic system mediates very diffuse widespread responses in the body, particularly involving the gut and the vascular system: for this reason, the system has come to be regarded as a mediator of massive blanket-like responses. The parasympathetic system, on the other hand, mediates more precise localized control of discrete parts of the body.

A more detailed description of the autonomic innervation of various vertebrate organs can be found in the general reviews of Burnstock (1970) and Campbell (1970). By far the most complex organ system of the body receiving an autonomic innervation is the alimentary canal. Much is now known about the details of its innervation and its reponses to autonomic activity and, since it receives both parasympathetic and sympathetic innervations, it forms an ideal model with which to illustrate the complexity of extrinsic autonomic control and the complexity of intrinsic nervous architecture.

The innervation of the mammalian gut

The various smooth muscles of the gut are innervated by both the sympathetic and the parasympathetic nervous system. The rough sketch in figure 2.2 shows the salient features of the main innervation pathways. The parasympathetic supply is in two main parts: (a) The vagus nerves to the oesophagus, stomach, small intestine and upper part of the colon.

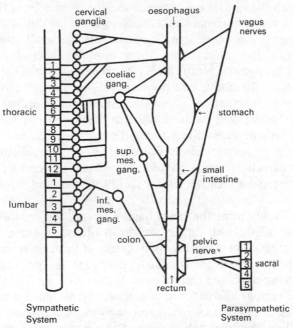

Figure 2.2. Scheme of innervation of mammalian gut by the sympathetic and parasympathetic systems. Sup. mes. gang. = superior mesenteric ganglion; inf. mes. gang. = inferior mesenteric ganglion.

(b) The pelvic nerve from the sacral segments 2–4 of the spinal cord to the lower half of the colon and the rectum. The vagal and pelvic preganglionic axons run almost directly to the innervated organs, synapsing directly with the postganglionic neurones of the gut wall plexuses.

The sympathetic innervation of the gut is far more diffuse. The oesophageal innervation comes from thoracic segments T1–T6, the postganglionic fibres passing from the superior, middle and inferior cervical ganglia. Some of the sympathetic fibres from the cervical ganglia join the vagus in its path to the oesophagus, while others pass from these ganglia direct to the oesophagus. In addition, some sympathetic postganglionic fibres run directly from the thoracic paravertebral ganglia to the oesophagus. The lower (abdominal) part of the oesophagus is supplied by sympathetic postganglionic fibres from the coeliac ganglion. The supply to the stomach comes mainly from thoracic segments T6–T9, to the small intestine from T9 and T10, and to the colon from T12–L2. The postganglionic fibres from these various thoracic and lumbar segments are routed through the various prevertebral ganglia. Fibres from the coeliac ganglion innervate the stomach and the top part of the duodenum, fibres from the superior mesenteric ganglion innervate the small intestine and the top part of the colon, while fibres from the inferior mesenteric ganglion and hypogastric plexus innervate the remainder of the gut.

In addition to the above efferent innervation, the gut possesses an extensive afferent innervation. These visceral afferent fibres pass back to the central nervous system via the nerves carrying the efferent sympathetic and parasympathetic fibres. Since these afferent fibres are not truly part of the two separate autonomic divisions, the terms 'sympathetic afferent' and 'parasympathetic afferents' should not be used. The correct terms for these fibres are *vagal afferents, pelvic* (or *sacral*) *afferents* and *thoracolumbar afferents*.

Visceral afferents from the oesophagus pass to the central nervous system via the vagus (with cells of origin in the nodose ganglion) and via the sympathetic fibres (with cells of origin in thoracic segments T5–T8). The afferents from the stomach also pass in the vagus while afferents from the colon and rectum pass via the pelvic nerves to segments S2–S4. These various afferents are stimulated by chemical action (as in heartburn via oesophageal afferents) and/or by mechanical action such as in gut distention. The operation of visceral afferents is, however, far more complex than this. An important point about much of their activity is that stimuli which normally activate afferents elsewhere in the body (i.e. pressure, heat, cold, etc.) are not experienced in the viscera. Many visceral afferents are stimulated by chemical and mechanical stimuli, giving rise to a variety of reflexes, but these do not rise to the level of consciousness. The main afferents running in sympathetic nerves run from the stomach and duodenum (to thoracic segments T7–T9) and from the rest of the gut to enter thoracic segments T9–T12. It appears that the sensation of pain in the gut is largely mediated via these thoracic afferents and not via vagal afferents.

The role of the afferent fibres in the physiology of the gut is to mediate the major inhibitory reflexes controlling gut motility. There is now much evidence, admirably reviewed by Youmans (1968), that afferent (i.e. sensory) fibres of enteric origin may send impulses via their axons in the mesenteric nerves which influence the discharge rate of the second (post-ganglionic) element of the afferent sympathetic pathways in the coeliac and mesenteric ganglia. This means that local reflexes will be possible due to sensory negative feedback directly onto postganglionic neurones without the necessary involvement of the central nervous system in the

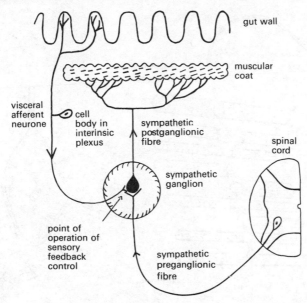

Figure 2.3. Conceptual scheme for the operation of sensory negative feedback for the control of intestinal motility (see text for details).

reflex. Figure 2.3 is a conceptual diagram illustrating the possible routing of just such a local inhibitory reflex. It must be kept in mind, however, that local reflexes of this type are still something of dispute in the literature.

Intrinsic nerve elements in visceral muscle

A considerable amount is now known about the neuroanatomy of the autonomic innervation of the visceral muscle. Much of the more recent microanatomical work, emanating from Professor Burnstock and his colleagues, has resulted from the application of fluorescence histochemical techniques. Some quite interesting contrasts with the effector axon/skeletal muscle innervation have been found, namely:

(1) Sympathetic axons run very long distances through the actual effector organ before terminating and the terminal system of neuromuscular junctions is very long.

(2) The neurones within the muscle have a prominent 'varicose' structure (i.e. they possess a string of swollen regions) with as many as 10–30 varicosities per 100 μm. The varicosity diameter may be up to 2 μm,

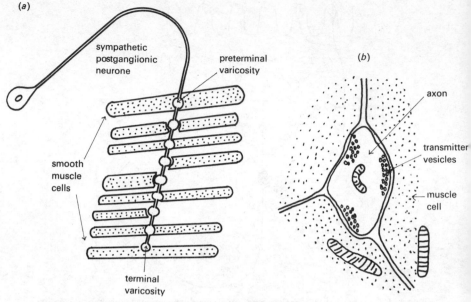

Figure 2.4(*a*). Scheme of a sympathetic/smooth-muscle motor unit involving the preterminal and terminal varicosities in end plate formation.

(*b*). Sketch showing the innervation of three smooth-muscle cells by a single axon (drawn from an electron micrograph in Burnstock, 1970).

while the intervaricose regions are only about 0·1–0·2 μm (see figure 2.4).

(3) Although the autonomic/effector neuromuscular junction is a relatively simple structure compared with that on most skeletal muscles (see Chapter 4), any one axon may give rise to very large numbers of such junctions (see figure 2.4).

(4) Transmitter concentration in the varicosities is high, and during a single impulse, vesicles are released from many varicosities.

This means that the autonomic axon makes a large number of 'en-passant' terminations on a large number of muscle cells, permitting many cells to be activated by a single impulse in the axon. In other words, a single efferent autonomic axon is responsible for the control of a true 'motor unit' (a group of dependent muscle cells). This situation is similar to the somatic nervous control of vertebrate skeletal muscles where the whole muscle is divided up into a series of separate motor units, each unit activated by a separate single axon.

There is considerable variation in both the innervation density and the extent of development of the intrinsic plexuses in various visceral muscles. In a work of this size it is not possible to discuss the neural micro-anatomy of the various viscera, and readers are referred to the excellent data collected together on this topic by Burnstock (1970), relating to plexus anatomy in the vas deferens, eye muscles, the male genital system, ureter, uterus, urinary tract and the vascular system. The gut is by far the most studied smooth-muscle system, and a brief account of its neural organization will suffice to illustrate the complexity of the terminal neural plexus.

The neural plexuses of the gut

The smooth-muscle coats of the gut are richly supplied with a complex laminated system of intrinsic plexuses. The sketch in figure 2.5 shows the typical organization of the enteric plexus system. The various sympathetic postganglionic fibres and parasympathetic preganglionic fibres sink into the gut wall, where their axons become inextricably interwoven in their path to the various ganglionated structures of the plexuses. There are three main plexus systems in the gut wall, the **subserous plexus**, the **myenteric plexus** (Auerbach's plexus) and the **submucous plexus** (Meissner's plexus). These plexuses are extensively interconnected by nerve fibres running centrifugally (i.e. from the outside to the inside of the gut) and centripetally (i.e. running from inner enteric neurones out from the gut).

The **subserous plexus** is the main connection point between the efferent sympathetic fibres and the rest of the gut plexus. As its name implies, it lies beneath the serosal gut layer. In intestinal muscle, this plexus is located at the attachment of the mesentery where the mesenteric nerve fibres enter the gut wall.

The **myenteric** (Auerbach) **plexus** is the main intramural plexus lying between the circular and longitudinal muscle coats, consisting of three plexiform networks. The primary plexus consists of large bundles of unmyelinated fibres linking a variety of ganglia of variable cell-body content. The secondary plexus is often called the interfasicular plexus; it consists of a group of small unmyelinated fibre bundles traversing the interstices of the primary plexus and containing only a few cell bodies. The tertiary plexus is continuous with the secondary plexus, consisting of fine nerve fibre strands. All three plexuses contribute unmyelinated fibres which ramify among the individual smooth-muscle cells of the gut longitudinal and circular coats.

The **submucous** (Meissner) **plexus** is extensively linked to the Auerbach plexus, both by inward-passing sympathetic fibres which synapse on the cell bodies of both plexuses, and by intrinsic neurones (see figure 2.5). The plexus consists of several meshworks of fibres and ganglia arranged in the submucosal layer, although the ganglia are small and contain few neurones. This plexus possesses fibres which pass further inwards to innervate the mucosa itself, for which the term *mucosal plexus* is sometimes given. However, this network contains no neurones and can best be considered as simply an extension of the Meissner plexus.

A great deal is now known about the light microscopical anatomy of the neurones of the various intramural plexuses, largely determined by intravital staining, and about the fine structure of these neurones by electron microscopy. Earlier workers distinguished the presence of three basic neurone types.

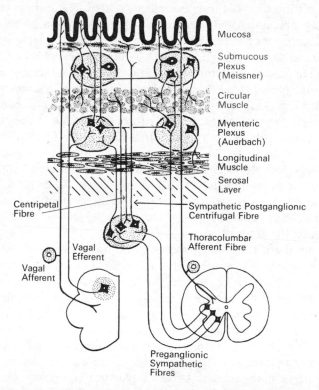

Figure 2.5. Scheme of plexus distribution and extrinsic innervation of the mammalian gut wall. See text for details. Modified and redrawn after Schofield (1968).

Type I neurones are characterized by numerous short highly-branched irregular dendrites, with a single long axon passing out of the ganglion. These were thought to be mainly motor in function.

Type II neurones possess far fewer dendrites which are long and smooth, with a complex arborizing axon confined to within the ganglion of origin. These neurones were thought to be largely sensory in function.

Type III neurones possess intermediate-length dendrites terminating in the ganglion of origin or in close neighbours, with a very long axon.

The neurones mentioned here are basically heteropolar (Types I and III) or isopolar (Type II) multipolar neurones. In more recent studies, however, isopolar bipolar neurones and unipolar neurones have been found.

The general neuronal anatomy of the intramural ganglia is far too detailed to be considered here. A most comprehensive account of this topic can be found in the review by Schofield (1968). Whereas earlier workers were of the opinion that the anatomy of the individual neurones gave some vital clues relating to their function, there is now considerable doubt as to the functional significance of the various neurone types. What is of importance, however, is the observation that most adrenergic neurones entering the gut end around these ganglion cells of the Auerbach and Meissner plexuses, and it is at this point that much of the modulation of gut activity occurs.

Visceral-muscle innervation in the invertebrates

Information about the nervous control of motility in invertebrate visceral muscles is extremely scanty, and much of it is confused and contradictory. By far the most important (and often the only really significant) visceral-muscle system in most invertebrates is the gut, but the gut itself is enormously variable in structure and activity in various phyla.

In the annelids, the gut is a simple muscular tube, but in polychaets, the anterior part, the proboscis, is reversible for grasping prey (as in most Nereids) or for ingesting detritus and sand (as in *Arenicola*). Little is known about the Hirudinea (Leeches), but in the oligochaets and polychaets which have been examined so far, all regions of the gut show peristalsis of some form, and there is evidence of co-ordinated neurogenic pacemaker activity to control gut motility.

In the molluscs, the stomach is highly contractile and most other regions of the gut show peristalsis. In its electrical activity and responses to KCl depolarization, the molluscan gut bears many resemblances to mammalian intestinal muscle.

In the echinoderms, the gut is highly contractile, and strong rhythmic peristalsis is usually present in both holothuria (sea cucumbers) and echinoids (sea urchins). Although single violent total gut contraction can occur, spontaneous rhythmic contractions constitute the normal level of activity (Prosser *et al.*, 1965), resembling mammalian intestinal activity.

In the arthropods, the gut consists of striated muscle fibres. In the insects and crustaceae, rhythmic peristalsis directs food down the foregut into the highly contractile stomach. In the midgut, alternating waves of peristalsis and antiperistalsis have been reported (i.e. contractions from anal to oral direction). Rhythmic peristalsis directs the gut contents into the first part of the hindgut, but in insects, irregular large non-rhythmic contractions are associated with pellet formation and expulsion. There is much evidence, reviewed by Prosser *et al.* (1965) of a complex neurogenic control of the arthropod gut, the intrinsic rhythmicity of the gut neurones being modulated by extrinsic control nerves. In some insects, however, the hind gut at least is not automatically active, activity being induced directly by extrinsic neurones.

The extrinsic control of the invertebrate gut

The innervation of the invertebrate gut is extremely varied due to structural variation within the various phyla, and to the extent of development of the nervous system. The concept of an enteric innervation in vertebrate terms, with functionally reciprocal sympathetic and para-sympathetic systems, is quite inappropriate since there is no evidence of any homology betwen the peripheral nerves in the two major divisions of the animal kingdom. Most studies have been carried out on annelids, arthropods, molluscs and echinoderms, and the great treatise by Bullock and Horridge (1965) considers innervation in detail in these groups. Here only some general observations can be made.

In annelids and arthropods, the term *stomodeal* or *stomatogastric* has been given to the gut innervation. This system (the nearest equivalent to the vertebrate autonomic system) connects to the somatic central nervous system, either directly at the brain or at anterior ganglia. In addition, the gut also receives an innervation in these long-bodied animals from the segmental ganglia of the ventral nerve cord. In annelids, the pharynx or oesophagus contain ganglia connected to the brain and to the circumoesophageal connectives (figure 2.6); motility of the muscle is caused by activity in these ganglia, the centres or route for strong central neural pacemaker control.

In earthworms, stimulation of the pharyngeal nerve supply causes gut inhibition, while the segmental supply to the gut from the ventral nerve cord contains both excitatory and inhibitory fibres. The excitatory fibres leave the central nervous system in the median and posterior nerves of each segment, their action being cholinergic. The inhibitory fibres leave the central nervous system in all three segmental nerves, their synaptic action being largely noradrenergic. However, noradrenaline may be both excitatory and inhibitory on different parts of the gut (as it is in different mammalian visceral muscles) and it is possible that the adrenoceptive system of invertebrate visceral muscle, like that of mammalian visceral muscle, may contain separate α and β-receptors (see Chapter 3).

In molluscs (figure 2.6, b and c) the gut is innervated anteriorly from the buccal ganglia and usually posteriorly from the visceral ganglion. Both buccal and visceral ganglion nerves to the gut are excitatory. In most molluscs, these excitatory nerves are cholinergic, and topically applied acetylcholine mimics their action. However, it is now known that in some molluscs (e.g. *Poneroplax*, *Venus* and *Tapes*), the hindgut may receive an inhibitory cholinergic innervation. Furthermore, there is also some evidence of adrenergic excitatory innervations in some molluscs. Adrenaline will induce contraction of the stomach and rectum of some cephalopods, and recent histochemical studies have shown the presence of monoaminergic nerves (most noradrenergic) in the hindgut muscle of *Helix* and the chiton *Poneroplax* (Campbell and Burnstock, 1968). These few examples show just how confused and uncertain is the whole picture of gut innervation in the molluscs.

In echinoderms, the gut is innervated from the circumoral nerve ring, the gut being inactive if the nerve supply is severed, suggestive of direct extrinsic control.

In the arthropods, the anterior part of the gut is innervated from the hindbrain, these nerves joining the intestinal plexus and extending back as far as the midgut. The mid and hindgut appear to be innervated from nerves emerging from the segmental ganglia of the ventral nerve cord, while the rectum receives an innervation from the last abdominal ganglion (figure 2.6d). The exact function of the extrinsic innervation is not clear. In crustaceae, ventral nerve cord stimulation either initiates or accelerates peristalsis, but when the posterior intestinal nerves are cut, peristalsis often ceases. Peristalsis may be initiated by direct gut stimulation. It is possible that peristalsis here may be neurogenic, its excitation arising in the cells

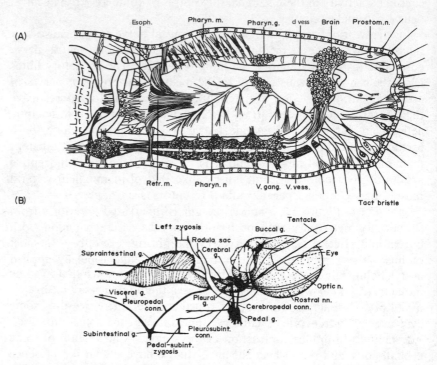

(A)

(B)

of the intrinsic plexus. If this is the case, then the extrinsic innervation must be simply modulatory in action, but it is difficult to explain why peristalsis ceases if the extrinsic supply is severed. It is clear from the literature that far more experimental studies will have to be carried out on invertebrate gut preparations before any interim analysis can be made of the importance or otherwise of extrinsic nerve control.

Intrinsic plexuses and the control of the gut

The structure and location of the enteric plexuses in a number of invertebrates has been reviewed by Bullock and Horridge (1965). In most cases, enteric plexuses, consisting of nerve fibres and cell bodies are located in the outer layers of the gut wall. When the gut coat is stratified into several layers, the plexus system is also stratified into outer and deeper layers. Where a very thin gut wall is present, as in many annelids, echinoderms and arthropods, the plexus runs along the cells of the intestinal epithelium at basal level.

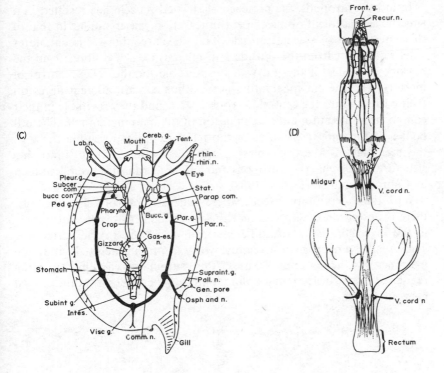

Figure 2.6. The innervation of the gut in some invertebrates.

(*a*). The anterior central nervous system and pharynx innervation in the annelid *Chaetogaster*. The pharyngeal ganglion produces two main nerves extensively innervating the muscle fibres of the pharynx on either side. Esoph., oesophagus; pharyn. g, m and n, pharyngeal ganglion, muscle and nerve; d vess., dorsal blood vessel; prostom. n., prostomial nerve; v. gang., ventral ganglion; retr. m., retractor muscle.

(*b*). Central nervous system and gut innervation of the prosobranch *Littorina*.

(*c*). Central nervous system and gut innervation of an opisthobranch; Bucc. g., buccal ganglion; gas-es. n., gastroesophageal nerve.

(*d*). Innervation of the gut of the beetle *Oryctes*. Front. g., frontal ganglia; recur. n., recurrent nerve. All adapted and redrawn from Bullock and Horridge (1965).

In most arthropods, the presence of striated muscle has resulted in a superficial placement of the efferent innervation rather similar to that of the innervation of skeletal muscle. The plexus contains efferent fibres entering the gut from the outside and afferent (sensory) fibres from the sensory neurones of the gut wall, responsible for the reflex control of motility. The efferent innervating nerves thus contain afferent fibres on their pathway into the central nervous system, and they resemble in their composition the autonomic innervation of the mammalian gut. The cell bodies of the enteric plexuses of most invertebrates resemble those of the mammalian Auerbach plexus, and their microanatomy and fine structure in polychaets, gastropods and amphineurans has been described by Campbell and Burnstock (1968).

With the exception of insects, where extrinsic control seems the rule, there is no reason to suppose that the neural control of visceral muscle in the invertebrates is any different from that seen in the vertebrate. This involves a neurogenic assembly, with excitation generated within the neurones of the enteric plexuses, this automatic rhythmicity being regulated by extrinsic excitatory and inhibitory neurones from the central nervous system.

CHAPTER 3

THE NEUROMUSCULAR JUNCTION AND TRANSMITTER SUBSTANCES

The autonomic neuromuscular junction

The application of electron-microscopic studies to visceral muscle in recent years has now established the nature of the autonomic neuromuscular junction. A series of detailed studies during the 1960s (for general review see Burnstock, 1970) have shown that in both vertebrate and invertebrate visceral muscles there is a close apposition of nerve and muscle membranes in most cases from 7–20 nm at the actual junction. Exceptions to this seem to be in many mammalian intestinal and vascular smooth muscles, where close apposition of pre- and postsynaptic elements is rare, the usual gap being 80–120 nm. A most unusual feature of autonomic axon terminations is the long series of nodular swellings or varicosities at preterminal and terminal positions (see Chapter 2). An interesting consistent observation is that the thin intervaricose regions are devoid of vesicles, the vesicles being confined to the varicosities. The implication here is that transmitter substances are liberated from a large number of sites along the axon, and that any one axon may make a series of neuromuscular junctions on one muscle cell or on a number of different muscle cells. The non-random distribution of vesicles leads us to the conclusion that one varicosity corresponds to one neuromuscular junction, a view confirmed by serial electron-microscopic sectioning. However, one varicosity may make junctions on more than one muscle fibre, this being particularly the case in vascular muscle.

An unusual feature of many mammalian intestinal and vascular smooth muscles is that the axons often remain within their bundles along with their accompanying glial cells, even when making synapses on muscle cells. The axon surface directly apposing the muscle cell is devoid of glial membranes, and the transmitter vesicles aggregate at this point. In these cases, there is usually a very wide neuromuscular synaptic gap of anything from

80 to 200 nm (see figure 3.1). Despite this very wide synaptic gap, it seems certain that these junctions still operate in the classic manner of presynaptic transmitter liberation, followed by diffusion to the receptor systems of the postsynaptic site. This method of synaptic activation may still be effective over gaps of up to 300 nm (Burnstock, 1970).

In its fine structure, the visceral neuromuscular junction is far less elaborate or specialized compared with the corresponding skeletal muscle structure, there being no equivalent of the end plate. The autonomic axon may make 'en passant' or terminal contacts with one or more muscle cells, and these contacts usually take the form of simple indentations or depressions of the cell surface. In some cases, quite deep burying of the axons may occur in very deep incursions (figure 3.1). However, even where deep incursion into the cell occurs, a clear synaptic gap is always present. There is surprisingly little evidence of any postsynaptic elaboration of the muscle-cell sarcolemma or sarcoplasm below the axon terminal (figure 3.2), there being no particular increase in mitochondrial

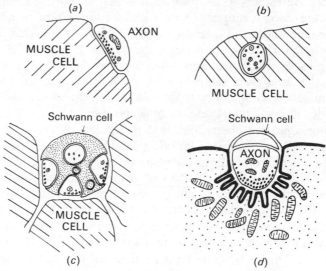

Figure 3.1 (*a* and *b*). The general morphology of the neuromuscular junction in molluscan and mammalian gut muscle. Notice that even when the axon is embedded within the muscle cell there is no particular elaboration of the postsynaptic site.

(*c*). Sketch of a bundle of axons making neuro-muscular synapses on a group of smooth-muscle cells of molluscan gut. Notice that Schwann-cell cytoplasm still partly covers the axons, and the neuromuscular gap is very wide (about 200 nm).

(*d*). Sketch of typical mammalian skeletal neuromuscular junction, showing the massive elaboration of the postsynaptic site, this being thrown into folds and containing a large aggregate of mitochondria.

or micropinocytotic vesicle populations in this region. This is a great contrast to the typical skeletal muscle elaborate end plate seen in the vertebrates (figure 3.1). In the smooth muscles of some parts of the vascular system and vas deferens, a few reports have shown a slight elaboration of subsurface cisternae (possibly a type of sarcoplasmic reticulum) at the postsynaptic site. This, however, is not a common observation, and its significance is not understood.

Various smooth muscles differ in the density of their innervation. In some muscles (e.g. rat and mouse vas deferens) each muscle cell receives an individual innervation, but this is clearly not the case in many other muscles, where activation may spread from cell to cell by tight nexal junctions. In addition, some invertebrate visceral muscles appear to be polyneuronally innervated (see figure 3.2), and this condition may well be the rule in the arthropods.

Transmitter vesicles

The fine structure of some mammalian visceral-muscle neuromuscular junctions is described in Burnstock (1970). Figure 3.2 shows the fine structure of some molluscan visceral-muscle neuromuscular junctions. What stands out in all of these cases, in direct contrast with skeletal-muscle junctions, is the very varied nature of the actual synaptic vesicles, the vesicles varying not just in size but also in appearance.

Three main types of synaptic vesicle have been found in both terminal and preterminal varicosities in the autonomic innervation of both vertebrate and invertebrate visceral muscle. These are shown in diagrammatic form in figure 3.3. Two obvious types of small vesicle are present, and a variety of types of large vesicle can be seen, all of a granular appearance. What is rather confusing is that more than one type of vesicle is usually present in any one presynaptic terminal.

Agranular small vesicles. These clear vesicles have a size range of 30 to 50 nm and they are identical with those found in skeletal-muscle neuromuscular junctions. The question naturally arises as to whether these vesicles represent the sites of acetylcholine storage, and several lines of investigation suggest that this is indeed the case.

(1) In preparations innervated by cholinergic nerves where the muscle cells respond to acetylcholine (e.g. bladder and ciliary muscles) these vesicles are the overwhelmingly predominant type.

(2) The neurones containing these vesicles in the Auerbach plexus can

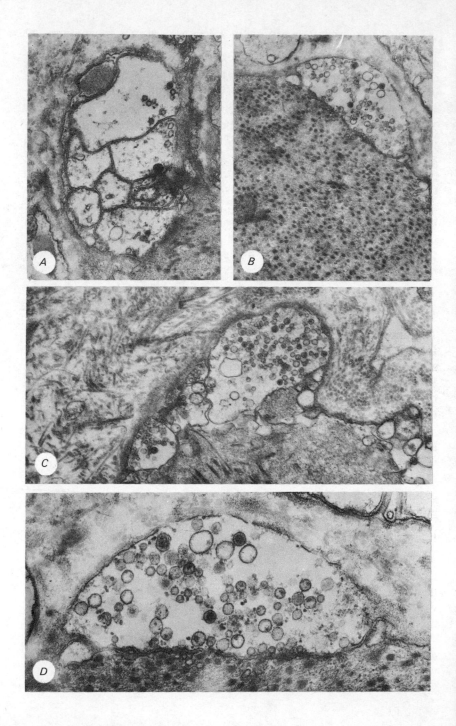

Figure 3.2. The fine structure of the molluscan-gut-muscle neuromuscular junction.
(*A*). Bundle of efferent axons, making synaptic contacts while still within the bundle. Print magnification × 23,000.
(*B*). Simple neuromuscular synapse on a single muscle cell. Print magnification × 16,700.
(*C*). Two axons making contact (exemplified by the vesicle content) with a muscle cell. Print magnification × 17,300.
(*D*). Enlargement of the synapse to show the different types of transmitter vesicle present. Print magnification × 33,000. Author's original plates.

be readily characterized by their cholinesterase histochemical staining properties.

(3) Neurones with this type of vesicle in predominance do not respond to adrenergic fluorescence staining techniques.

In the presynaptic sites where these vesicles are found, they represent the great bulk of the vesicles present, but some large granular vesicles are almost invariably present.

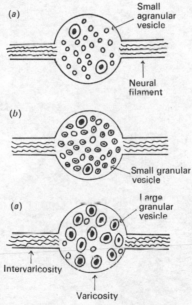

Figure 3.3. Types of vesicle found in typical autonomic neuromuscular junctions on smooth muscle.
(*a*). Typical vesicle composition of parasympathetic cholinergic neurones, with a predominance of small agranular vesicles and some large granular vesicles.
(*b*). Typical vesicle composition of sympathetic noradrenergic neurones, with a predominance of small granular vesicles, plus some small agranular vesicles and some large granular vesicles.
(*c*). Typical vesicle composition of many neurones of the vertebrate gut, and the amphibian bladder and lung. Adapted and redrawn from Burnstock (1970).

Granular small vesicles. These have a size range of 30–50 nm and they predominate in sympathetic neurones, suggesting that they may be the sites of noradrenaline storage. That this is indeed the case can be deduced from the following evidence:

(1) These vesicles predominate in sympathetic neurones which have been shown by fluorescence histochemistry to contain very high levels of noradrenaline in the varicose regions (figure 3.4).

(2) In the developing mouse vas deferens, as there is an increase in the number of identified noradrenergic nerves (shown by histochemical methods), there is a parallel increase in the number of neurones containing small granular vesicles (Yamaguchi and Burnstock, 1969).

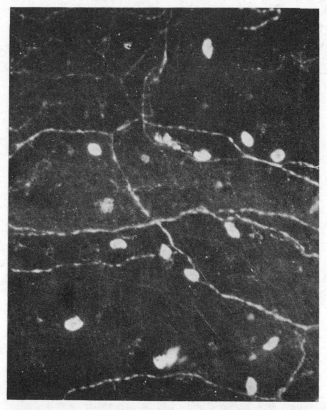

Figure 3.4. Fluorescence histochemical localization of monoaminergic neurones (almost certainly noradrenaline) in rat iris. Notice the beaded (varicose) appearance of the neurones. The very large white cells are mast cells. Print magnification × 500. Print courtesy of Dr. D. P. Knight.

(3) Autoradiographic electron-microscopic studies of sympathetic nerves injected with tritiated noradrenaline show silver-grain development (i.e. the loci of noradrenaline accumulation) in the regions of small granular vesicle accumulation (Burnstock, 1970).

(4) Drugs such as reserpine, which are known to deplete noradrenaline stores, remove the granules from these vesicles when the sympathetic nerves are exposed to the drugs.

(5) Monoamine oxidase inhibitors, when administered along with noradrenaline or related substances such as dopamine, produce increases in the small granular vesicle population of sympathetic neurones.

(6) Noradrenaline has been conclusively localized by electron-microscopic histochemistry in the cores of the small granular vesicles of rat vas deferens neurones.

In the presynaptic sites where these vesicles dominate, some large granular vesicles are also present. However, an additional complication to this picture is the presence also of some small clear agranular vesicles at these sites. In the mouse vas deferens, Yamaguchi and Burnstock (1969) estimated the vesicle distribution as about 85% small granular vesicles, 12% small agranular vesicles, and 3% large granular vesicles. Several possibilities immediately present themselves:

(a) The small agranular vesicles contain acetylcholine, hence the site must release both excitatory and inhibitory transmitter.

(b) These vesicles contain an as yet unidentified transmitter which could be excitatory or inhibitory.

(c) These vesicles are simply empty noradrenaline-depleted vesicles in the process of transmitter re-accumulation, and the site releases only one main transmitter, noradrenaline.

Two main lines of evidence suggest that these vesicles are merely empty noradrenergic granular vesicles. Firstly, when drug treatments are used to increase the neuronal noradrenaline store (such as the use of monoamine oxidase inhibitors), the number of agranular vesicles falls and a corresponding increase occurs in the proportion of small granular vesicles. This suggests that these small agranular vesicles have the ability to store newly synthesized noradrenaline. Secondly, it is known that after transmitter release, the vesicle membrane joins the presynaptic terminal membrane and is later retrieved by a process of micropinocytosis (Smith, 1970; Huddart and Bradbury, 1972). These vesicles readily take up 5-hydroxydopamine under experimental conditions and develop dense cores, a further indication that these vesicles are noradrenergic. A

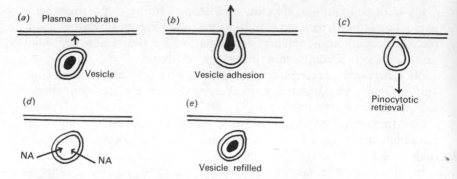

Figure 3.5. Hypothetical scheme showing the possible fate of the small granular vesicle after liberation of its contents into the synaptic cleft. This may afford an explanation of the presence of small agranular vesicles in sympathetic noradrenergic neurones.

conceptual diagram of how this may occur is shown in figure 3.5. The small fraction of the total vesicle population which is agranular (about 12%) may thus be that relatively constant fraction which at any one time is restoring noradrenaline. However, we cannot rule out the possibility that these small agranular vesicles may contain some other transmitter substance (such as 5-hydroxytryptamine or dopamine), but there is insufficient information on this point to be any more certain at the moment. What must be kept in mind is that some small agranular vesicles in some nerves are neither adrenergic or cholinergic (Daniel, 1973).

Granular large vesicles. These are sometimes called *neurosecretory* vesicles and they have a size range 70–160 nm, with a central core varying in density and size (39–90 nm). In some neurones, these vesicles predominate, e.g. in some blood vessels and in the nerves to the amphibian lung and bladder, while some axons to the toad intestine contain only large granular vesicles. Furthermore these vesicles predominate in the cell bodies of neurones whose terminal varicosities contain small granular vesicles. Much of the variation in the literature concerning vesicle distribution may be due simply to selective electron-microscopic section sampling, or differing fixation techniques. It seems probable that large granular vesicles always occur in axons along with small vesicles, but they may have different loci of aggregation. Some of these problems are discussed by Burnstock (1970).

The function of these vesicles is something of a mystery at the moment. In addition to cholinergic excitatory and adrenergic inhibitory fibres in the gut wall, there are also some non-adrenergic inhibitory fibres present. These show neither cholinesterase-positive histochemical reactions nor noradrenergic fluorescence reactions, and Daniel (1973) has suggested that these neurones may be those in the gut containing predominantly large granular vesicles. There is clear evidence that non-adrenergic inhibition in the gut is blocked by nicotinic and 5-hydroxytryptamine blocking agents, suggesting that some non-adrenergic inhibitory neurones may release acetylcholine, while some definitely release 5-hydroxytryptamine. Since the large granular vesicles are non-cholinergic, it is possible that they may contain 5-hydroxytryptamine, which is an established autonomic transmitter.

Since the large granular vesicles vary in size and appearance, it seems likely that they may play many different roles. There is certainly evidence that some large granular vesicles may contain some monoamines such as adrenaline. This appears to be the case in some amphibian sympathetic neurones, although reserpine-induced adrenaline depletion does not seem to affect large granular vesicles, and it also seems to be the case in adrenal medulla vesicles. There is also evidence for association of these vesicles with 5-hydroxytryptamine in molluscan neurones and in the rat pineal body.

It is possible that the large granular vesicles in mammalian sympathetic neurones may be the loci of transmitter synthesis, while the small granular vesicles are the loci of transmitter storage prior to release. Certainly noradrenaline is synthesized in the cell body and here the large granular vesicles predominate, the noradrenaline being translocated down the axon to the terminals. The large granular vesicles in the terminal region may simply represent additional 'on site' loci of noradrenaline synthesis. However, this does not help to explain the function of the large granular vesicles in cholinergic neurones where no noradrenaline histochemical localization can be demonstrated. Some experimental evidence has pointed to the accumulation of dopamine in large granular vesicles (a possible additional sympathetic transmitter), but not in those of cholinergic endings.

Clearly it is possible in some cases to point to an involvement of large granular vesicles in transmitter activity, but in many cases no such involvement can be shown (as in parasympathetic cholinergic neurones). The possibility arises that the large granular vesicles in many cases may have no transmitter activity, but may be related to some kind of trophic activity. One piece of evidence suggesting this is that the large granular vesicle population progressively decreases during the development of the

Table 3.1.—Principal neuromuscular transmitters and their modes of action on visceral muscle.

Transmitter	Major sites of action
ACETYLCHOLINE	Main parasympathetic transmitter, excitatory on the smooth muscle of the alimentary canal but inhibitory action on vascular smooth muscle.
NORADRENALINE	Major inhibitory transmitter on the smooth muscles of the gut. The excitatory transmitter over much of the vascular smooth musculature.
ADRENALINE	Major inhibitory transmitter in the sympathetic innervation of amphibian gut smooth muscle.
5-HYDROXYTRYPTAMINE	May be the inhibitory transmitter of some mammalian gut neurones, and in the molluscan gut.
DOPAMINE	A possible sympathetic transmitter in mammals.
GLUTAMIC ACID	The established neuromuscular transmitter of arthropod skeletal muscle, possibly including the striated muscle of the gut.

mouse vas deferens, reaching a low equilibrium point when the tissue is mature (Yamaguchi and Burnstock, 1969). The trophic influence of neurones on muscle cells is an established phenomenon in skeletal muscle (Guth, 1968). Denervated muscle fibres show many profound changes in both the sensitivity of the postsynaptic site to transmitters, and in the specificity and distribution of the receptor mechanism on the cell surface. It seems possible that the nerve terminal in both skeletal and visceral muscle may maintain the normal functioning of the postsynaptic site transmitter receptors by the release of chemical agents; and it is possible that some at least of the large granular vesicles may be involved in this trophic maintenance (see Osborne *et al.*, 1971; Atwood *et al.*, 1971; Huddart and Bradbury, 1972). However, at the present level of our understanding of transmitter physiology, this view must be regarded as highly speculative.

Transmitter substances

The main transmitter substances implicated in activation or inhibition of visceral muscles are shown in Table 3.1. Of these substances, by far the best-documented are acetylcholine and noradrenaline, and much is now known about their mode of action at the postsynaptic site.

Acetylcholine

Acetylcholine, which is a simple quaternary ammonium compound, is of widespread occurrence in plants and animals. Since it forms the main chemical transmitter of the nervous system of all main animal phyla, it must be regarded phylogenetically as the most ancient archetypal transmitter agent. Acetylcholine is manufactured by the transfer of an acetyl radical from acetyl co-enzyme A to choline by the choline acetyltransferase enzyme system:

$$CoA-S-\overset{\overset{\textstyle O}{\|}}{C}-CH_3 + HO-CH_2-CH_2-\overset{+}{N}(CH_3)_3 \rightarrow CoA-SH +$$
$$\overset{\overset{\textstyle O}{\|}}{C}H_3-C-O-CH_2-CH_2-\overset{+}{N}(CH_3)_3$$

This enzyme system is present in the presynaptic terminal cytoplasm where the acetylcholine itself is produced prior to storage in the vesicles. Each vesicle is thought to contain several thousand acetylcholine molecules. After release into the neuromuscular junction, the acetylcholine molecules diffuse to the postsynaptic site where they interact with specific cholinoreceptors, this reaction resulting in a sharp increase in the permeability of the postsynaptic site to ions. The cholinoreceptor mechanism reacts in such a way to acetylcholine that its conformation changes, resulting in the opening of channels in the membrane through which ions can pass. This activity, which occurs only in the synaptic region, lasts only a few milliseconds, and it is terminated by the activity of acetylcholinesterase, which hydrolyses acetylcholine to choline and acetic acid. The choline is reabsorbed by the presynaptic terminal for acetylcholine resynthesis. Excellent detailed accounts of the cholinergic synapse can be found in Eccles (1965) and Michelson and Zeimal (1973).

Acetylcholine/cholinoreceptor interaction

Belleau (1964) advanced the view that acetylcholine and certain other agents now called *cholinomimetics* induce specific conformational changes in the cholinoreceptor molecule to increase membrane permeability to ions. Acetylcholine blocking agents, on the other hand, induce non-specific conformational changes in the cholinoreceptor, due to the formation of hydrophobic bonds which do not alter the membrane permeability to ions. Although it has not yet proved possible to isolate individual substances which are unequivocally cholinoreceptive agents, an immense amount of work is in progress on this problem. The actual cholinoreceptor is known to be a water-soluble protein

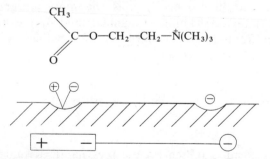

Figure 3.6. Conceptual diagram of the interaction of acetylcholine with the postsynaptic cholinoreceptor molecule, showing the dipole-dipole esteric centre and the anionic centre. From the point of view of charge distribution at the active centres, the receptor can be represented as in the lower figure.

present on the outer lipoprotein layer, and the structure of the active surface of the receptor is known to be related to the actual structure of acetylcholine.

The acetylcholine molecule possesses two active groups, a trimethyl-ammonium group with a positive change, and an ester group which is very strongly polarized. It is thought that there are two complementary groups on the cholinoreceptor molecule. One group, the *anionic* group, forms an ionic bond with the cationic head of the acetylcholine molecule. The other group is thought to make a dipole-dipole interaction (figure 3.6) with the esteric group of the acetylcholine molecule. There is over-whelming evidence that it is the anionic site which plays the main role in the cholinoreceptor interaction. The tetramethylammonium molecule, which possesses *only* a cationic group and can thus interact only with the anionic site of the cholinoreceptor, possesses very strong cholinomimetic activity. In addition, if the trimethylammonium group on acetylcholine is replaced by one without the charge (i.e. unable to interact with the

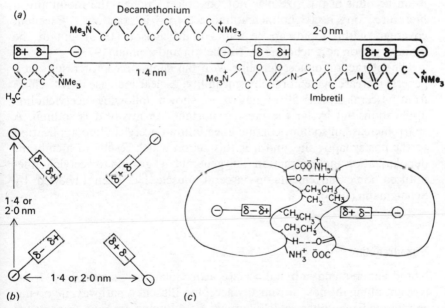

Figure 3.7. Possible schemes of mutual distribution of cholinoreceptors at the postsynaptic site.
(*a*). The linear scheme, showing distances between the anionic sites based on a C10 distance (as in decamethonium) and on a C14 scheme (as in imbretil).
(*b*). The tetrameric scheme, with either a 1·4 or 2·0 nm distance between all four anionic sites.
(*c*). The dimeric scheme. Adapted and redrawn after Michelson (1973).

anionic cholinoreceptor site), cholinomimetic activity falls to negligible levels.

An important aspect of cholinoreceptor action is the actual disposition of the individual receptors on the postsynaptic surface. The intense cholinoreceptor blocking activity of decamethonium (figure 3.7) led investigators to propose that cholinoreceptors are located in such a way that the distance between their anionic sites is equal to the internitrogen distance of the decamethonium molecule (about 1·4 nm). This is often called the 10C structure disposition. In more recent times, the synthesis of even more powerful cholinoreceptor blocking agents with 14 and not 10 carbon atoms in the internitrogen chain has suggested that the interanionic cholinoreceptor site distance may be nearer to 2 nm (known as the 14C structure disposition). The receptors are not scattered in a random fashion, but are aggregated into oligomeric complexes. The exact complex arrangement may well differ in different cholinergic synapses, and some proposed disposition schemes are shown in figure 3.7. Clearly in an account of this size it is not possible to review the mountain of literature on acetylcholine/cholinoreceptor interactions. Excellent coverage of this topic can be found in Michelson (1973) and in the monograph on acetylcholine by Michelson and Zeimal (1973).

The interaction of acetylcholine with the cholinoreceptor results in a general increase in membrane permeability. A clear increase in K^{42} efflux from visceral-muscle fibres has been shown following acetylcholine application, but by far the most important ion involved is sodium. A sharp increase in sodium conductance follows acetylcholine application to the postsynaptic site, and it is this action which results in membrane depolarization and contraction. The general background to acetylcholine-induced ionic movements in visceral muscle has been reviewed by Schatzmann (1968).

Noradrenaline

Noradrenaline and adrenaline are catecholamines synthesized from tyrosine along the now universally accepted Blaschko pathway, involving as intermediates other catecholamines now known to have transmitter activity. The discovery of the enzyme Dopa decarboxylase made it possible to propose a simple synthesis pathway

TYROSINE—DOPA—DOPAMINE—NORADRENALINE—ADRENALINE

as follows:

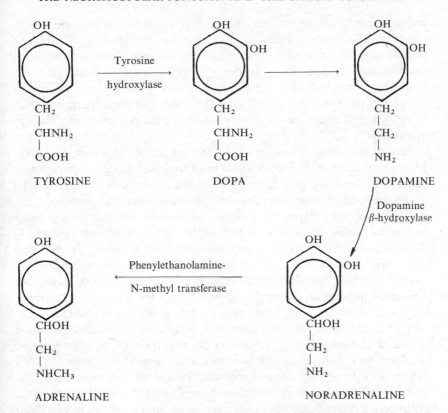

Although the above is regarded as the major pathway, the broad specificity of so many of the enzymes involved makes several minor alternatives possible, and the varying pathways may be used to a greater or lesser degree by different animal groups.

Noradrenaline/adrenoreceptor interaction

It is now known that there are two quite different types of adrenergic receptor, activation of which results in two quite different kinds of visceral-muscle response to adrenergic substances. Vasoconstriction, uterine contraction, uretal contraction and intestinal relaxation result from relatively low concentrations of adrenaline and noradrenaline, and these actions are mediated by catecholamine interaction with alpha receptors. Uterine relaxation and dilation of the vascular system is associated with catecholamine interactions with beta receptors. Generally, with the exception of the gut, alpha effects are excitatory while beta effects are inhibitory. There is now evidence that intestinal smooth muscle may contain both alpha and beta receptors, and activation of either receptor leads to relaxation. There is also evidence that the inhibitory effect mediated by alpha receptors is through an action on the neural elements in Auerbach's plexus, and that catecholamines may directly inhibit some muscle fibres through beta receptors in the fibre membrane.

Remarkably little is known about the nature of the adrenoreceptors. Increase in the size of the cationic head makes a catecholamine more active on a beta receptor than on an alpha receptor. This may not be of any real significance since any one catecholamine, e.g. noradrenaline, can activate both types of receptor, the effect on the organ being dependent upon the nature of the organ. What is interesting here is that the two different kinds of receptor, although activated by the same substance, produce opposite effects in different tissues, e.g. being excitatory on the vascular system and inhibitory on intestinal muscle. This must mean that, although the adrenoceptors are not very specific, the muscle-fibre-membrane response to their activation is specific. In the case of both adrenergic excitation and adrenergic inhibition, the main action at the postsynaptic surface is an alteration of membrane ionic conductances. Where catecholamine action results in inhibition, this action may be caused by a hyperpolarization of the membrane; the increase in potassium permeability brings the membrane nearer to a potassium electrode potential or it may be by an increase in intracellular calcium binding. In the cases where noradrenaline causes excitation (e.g. vasoconstriction), this is presumably caused by increase in sodium or calcium permeability. However, definitive work on this topic is lacking.

CHAPTER 4

THE ELECTRICAL ACTIVITY OF
VISCERAL MUSCLE AND ITS IONIC BASIS

The cycle of events resulting in the contraction of a muscle fibre is triggered by a sharp and pronounced depolarizing change in the surface-membrane polarity. Since the activation mechanism of muscle is electrical in nature, our consideration of visceral-muscle function should thus logically begin with an examination of the electrical characteristics and activity of the muscle-fibre membrane.

The resting potential

Although the presence of a large internal negative resting membrane potential is a characteristic of excitable tissues, the experimental measurement of the resting potential of visceral muscle is far from an easy task. Visceral-muscle cells, which are either irregular in shape or spindle-like and in the 2–5 μm diameter range, are very much smaller than skeletal muscle fibres (usually in the 30–100 μm diameter range), and this makes impalement with conventional glass capillary intracellular micro-electrodes of 0·5 μm tip somewhat unpredictable. (For this reason, higher-resistance electrodes with smaller tips are used.) Not only is the risk of cellular damage high during the initial impalement, with a resulting decline of recorded membrane potentials, but the inherent autorhythmicity of so many single-unit muscles simply acts to accentuate this mechanical damage.

The presence of autorhythmicity also adds a complication to the actual definition of what we mean by 'resting potential'. Many visceral muscles are spontaneously active and their resting membrane potentials, far from being maintained at a stable level (as is the case in multi-unit muscles and skeletal muscle) are rhythmically interrupted by slow wave-like depolarizations upon which there are usually superimposed a variety of

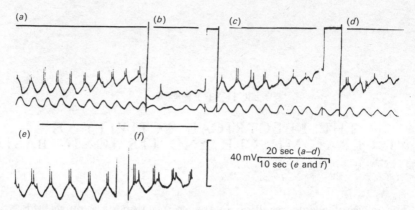

Figure 4.1. Slow waves and spikes recorded with flexibly mounted KCl intracellular electrodes from cells of the rat ileum. Each sequence of traces from (a) to (f) represents the impalement of a separate cell, the upper line representing zero potential, the middle line the membrane potential, and the lower trace the mechanical response of the tissue.
(a), (b), (c), (d) and (f) are records from cells damaged during impalement, and a gradual fall in membrane potential is observed, with a reduction of the slow waves and spikes.
(e) shows what can be regarded as an ideal record. Note the variation in spike numbers per slow wave and variation in slow-wave size and rate of depolarization. Calibrations: 40 mV and 20 sec (a–d) and 10 sec (e and f). Records courtesy of Dr. A. J. Syson.

spike-like discharges (see figure 4.1). In the case of these muscles, it is usual to define the resting potential as the point of maximum polarity, i.e. at the trough between the slow waves where membrane negativity is greatest.

Tables 4.1 and 4.2 show resting and action potential values of some representative rhythmic and non-rhythmic visceral muscles. Two trends emerge from these data. Firstly it is clear that the resting potential of these visceral muscles is rather low compared to comparable data from cardiac and skeletal muscle; and secondly it is also clear that the resting potential of rhythmic muscle is consistently lower than that of the multi-unit non-rhythmic type.

The action potential

The action potential of visceral muscle is exceedingly variable both in shape and duration, a striking contrast with the relative uniformity of this event in skeletal muscle, and a large number of different action potential 'types' have been described. In their excellent review, Burnstock *et al.* (1963) have placed in perspective much of the older work on action potential studies, and it is clear that much of this descriptive electro-

Table 4.1.—Membrane potential characteristics of some representative non-rhythmic (multi-unit) visceral muscles.

Animal and muscle	Resting potential (mV)	Action potential (mV)	Action potential duration (msec)	Reference
frog tongue artery	− 64·7	60	100–200	Steedman (1966)
frog abdominal mesenteric artery	− 43·6	—	—	Steedman (1966)
guinea-pig vas deferens	− 60	80	20–30	Bennett (1967)
guinea-pig uterine artery	− 60·7	c. 60	1,000	Bell (1969)
guinea-pig ureter	− 60	66	100	Washizu (1966)
rabbit common carotid artery	− 44·5	15·7	1,500	Mekata (1971)
chick gizzard		c. 8	500	Bennett (1970)
pigeon gizzard	—	10–12	500	Bennett (1970)
cat ureter	− 42	44	150	Kobayashi (1969)
rat uterus	− 40	40	100	Anderson *et al.* (1971)
rat ureter	− 57	68	—	Burnstock and Prosser (1960*a*)
dog retractor penis	− 68	68	—	Burnstock and Prosser (1960*b*)
squid chromatophore muscle fibres	− 20 to − 50	c. 20	200	Florey and Kriebel (1969)
cockroach rectum	− 48	35–40	500	Nagai and Brown (1969)
cockroach rectum	− 30	10	150	Belton and Brown (1969)
chiton rectum	− 63	—	—	Burnstock *et al.* (1967)

physiology is of little real meaning now. It is unfortunate that so many early investigations paid little attention to exactly what action-potential parameters were important in terms of the contraction of the cell. The consensus view from such studies was that contraction of the visceral-muscle fibre was almost invariably associated with spike-like depolarizations of the membrane potential, these spikes constituting part of what could be a very complex action-potential waveform. An apparent anomaly in early studies was the observation that contraction of visceral-muscle fibres could sometimes take place without spike-like action potentials, the only recordable electrical event being a slow sinusoidal wave.

Advances in our understanding of the excitation-contraction coupling mechanism (see Chapter 5) now make it clear that a spike is not necessarily a prerequisite for contraction in any type of muscle. As long as some part of the action-potential waveform depolarizes the membrane beyond the mechanical threshold level for the engagement (triggering) of

Table 4.2.—Membrane potential characteristics of some representative rhythmic (single-unit) visceral muscles.

Animal and muscle	Resting potential (mV)	Action potential (mV)	Action potential duration (msec)	Reference
guinea-pig taenia coli	−57	55–60	80–100	Casteels and Kuriyama (1966)
rat mesenteric blood vessel	−39·4	c. 20	150	Steedman (1966)
guinea-pig portal vein	−37	c. 40	200–500	Ito and Kuriyama (1971a)
carp stomach	—	c. 20	2,000	Ito and Kuriyama (1971b)
cat duodenum	—	c. 14	3,000	Kobayashi et al. (1967)
cat intestine	−52	38	1,500	Burnstock and Prosser (1960)
rat ileum	−40·7	18–25	2,500	Syson and Huddart (1973)
chiton intestine	−38	—	—	Burnstock et al. (1967)
frog abdominal skin blood vessel	−26	36	c. 2,000	Funaki (1961)
rabbit taenia coli	−25 to −50	5–20	100	Daniel and Singh (1958)
rabbit colon	−49·6	33·5	—	Gillespie (1962)

the excitation-contraction coupling mechanism, then the fibre will contract. This depolarization could be a spike or a slow wave, or it could be caused by an applied artificial square pulse or an increase in external potassium. So little work has been done on mechanical thresholds of visceral muscle that it is not possible to say whether distinct thresholds for contraction occur as is the case in skeletal muscle (Hodgkin and Horowicz, 1960). In rat ileal muscle, Syson and Huddart (1973) failed to establish a sharp membrane potential point at which tension was suddenly generated. Instead, an almost linear relation was seen between tension and membrane potential, indicating that the spike itself is not of fundamental importance.

Whether the primary depolarization of the plasma membrane be by a spike or a slow wave is probably not of great significance, but what is of significance is the duration of the action potential event. Long plateau-like depolarizations such as those seen in the rabbit common carotid artery, or cockroach rectum, and multiply-grouped spikes such as those of the guinea-pig ureter and chick mesenteric artery prolong the membrane potential depolarization, thus prolonging the active state of the muscle and thereby the duration of the mechanical contraction. By contrast, brief spikes such as those seen in the guinea-pig taenia coli, vas deferens, and in the rat ileum produce briefer and sharper periods of tension generation, in some ways analogous with slow skeletal muscle twitches. What emerges from this is that the electrical activation process of visceral muscle is just as varied as the mechanical output, this in turn being as varied as the functions which visceral muscle is called upon to perform in its various locations within the body. An additional complication is that unusual multiple or prolonged spike-like events may be caused by electrotonic spread from adjacent cells.

Tables 4.1 and 4.2 give action potential data from a variety of visceral

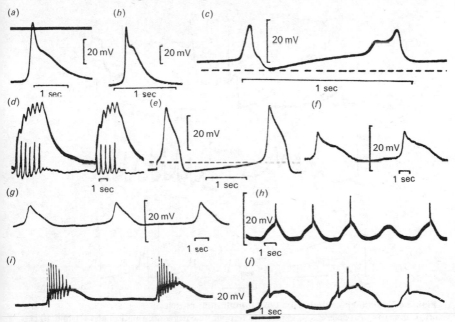

Figure 4.2. Action potential records from typical single-unit visceral muscles.
(a). Guinea-pig portal vein (from Ito and Kuriyama, 1971a).
(b). Guinea-pig taenia coli (from Casteels and Kuriyama, 1966).
(c). Rat mesenteric arteriole (from Steedman, 1966).
(d). Chick anterior mesenteric artery (from Bolton, 1968).
(e). Action potentials from the cockroach rectum firing from a spontaneous postsynaptic potential (first record) and a pacemaker-like depolarization (second record) (from Nagai and Brown, 1969).
(f). Rabbit common carotid artery (from Mekata, 1971).
(g). Carp stomach muscle (from Ito and Kuriyama, 1971b).
(h). Cat duodenum (from Kobayashi et al., 1967).
(i). Rabbit small intestine showing action potentials with many and few spikes per wave (from Gonella, 1965). Calibrations in all traces, 20 mV and 1 sec.

muscles, and records from some representative muscle types are shown in figures 4.2 and 4.3. These action potentials have been arbitrarily divided into spontaneous-action potentials and non-spontaneous-action potentials, the latter being evoked either by nerve stimulation or by direct stimulation of the muscle. However, it must be borne in mind that in some preparations such as the cockroach rectum, both spontaneous-action potentials and excitatory-junction potentials from nerve stimulation can be found. As a major generalization, it is evident that in non-spontaneous muscles the resting potential is higher, and the spike has a greater tendency to overshoot zero potential than is the case in rhythmically-

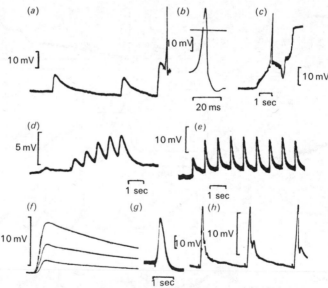

Figure 4.3. Action potentials from typical multi-unit muscles.

(a). Guinea-pig vas deferens, e.p.s.ps. from hypogastric nerve stimulation at 1 Hz and 10 Hz, producing a spike-like response (from Ferry, 1967).

(b). Guinea-pig vas deferens spike resulting from a single stimulus to the hypogastric nerve (from Bennett, 1967).

(c). Guinea-pig uterine artery, action potentials from perivascular nerve stimulation at 20 Hz (from Bell, 1969).

(d). Guinea-pig uterine artery, e.p.s.ps. from perivascular nerve stimulation at 2 Hz (from Bell, 1969).

(e). Chiton rectum, e.p.s.ps. from direct stimulation at 2 Hz (from Burnstock et al., 1967).

(f). Cockroach rectum, e.p.s.ps. from stimulation of proctodeal branch of cercal nerve at three different intensities (from Nagai and Brown, 1969).

(h). Squid chromatophore muscle fibres, e.p.s.ps. (from Florey and Kriebel, 1969).

Calibrations: (a) 10 mV; (b) 10 mV and 20 msec; (c) 10 mV and 1 sec; (d) 5 mV and 1 sec; (f) 10 mV; (g) 10 mV and 1 sec; (h) 10 mV.

active muscle. There are, however, some exceptions to this rule, such as the guinea-pig taenia coli and portal vein. What does stand out, however, is that in most intestinal muscles the spike greatly undershoots zero potential.

The action-potential responses of visceral muscles are far more variable in time course than those of cardiac and skeletal muscles, and they can be roughly divided into plateau-like and spike-like types. These differ in duration from brief (40 msec) spikes to long (2–15 sec) plateaux, but there is great confusion surrounding exactly what is meant by 'plateau-like' action potentials in visceral muscle. Immediately what springs to mind is the type of action potential characteristic of cardiac muscle, and action potentials of this type have been described in ureteral muscle of guinea-pig and rat.

Here the action potential consists of an initial fast spike whose repolarization phase is delayed and prolonged. However, more modern work on guinea-pig ureter by Washizu (1966, 1967, 1968) has shown that the plateau in reality consists of a series of spikes during which the membrane potential oscillates from about $+10$ to -30 mV. Some preparations which normally respond with a spike to neural stimulation give plateau action potentials when stimulated with long-lasting intracellular pulses (Bennet, 1967). However, many cases of naturally occurring plateau-like action potentials are known. The rabbit common carotid artery has classic plateau action potentials of about 2 msec duration; while those of the turtle aorta may last up to 15 sec. In all cases, it is the delayed repolarization which makes up the bulk of the plateau time course.

Spike-like action potentials show great variability. Spikes may occur both spontaneously, as in guinea-pig portal vein, or in response to electrical stimulation, potassium depolarization or stretch, as in the guinea-pig vas deferens (Bennett, 1967) and uterine artery (Bell, 1969). Some spikes rise abruptly, in an all-or-none fashion, while others may be graded. In some cases, the graded nature of the spikes may be related to the number of spikes in a single burst, or to the actual resting potential value of the cell. An excellent example of spike variability is shown by the guinea-pig ureter. When devoid of the renal end pacemaker, spikes may be elicited by direct electrical stimulation. In many cells, the action potentials significantly overshoot zero potential (figure 4.4a), but this is invariably correlated with high cell resting potentials. In cells with low resting potentials, even the multiple-spike discharges rarely overshoot or even achieve zero potential (figure 4.4b). This

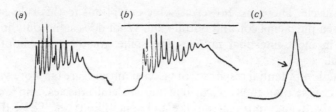

Figure 4.4. Spike discharges during the action potential of guinea-pig ureter.
(a) Action potential with multiple spikes and strong overshoot of zero potential in a high resting potential cell ($-65\,$mV).
(b). Multiple spikes not overshooting in a cell with low resting potential ($-50\,$mV).
(c). Single spike in a cell with low resting potential ($-52\,$mV). From Washizu (1966).

particular preparation is interesting in that the number of spikes in the action-potential response to stimulation is quite variable. Single-spike action potentials almost invariably fail to overshoot zero potential (figure 4.4c), and this type of response is more commonly found in cells with low resting potentials. Where multiple-spike action potentials occur, considerable variation in individual spike height is seen in both low and high resting-potential cells. The spike here is clearly not an all-or-none event; graded electrogenesis would appear to operate.

The slow wave

A major variation among visceral muscles with spike-like action potentials exists in the presence or absence of slow waves. In some muscles, such as the guinea-pig ureter, the spike rises abruptly from the resting potential level, but in many visceral muscles, such as those of the alimentary canal and some parts of the vascular system, the spike is found superimposed at various points upon a slow wave-like spontaneous depolarization. This type of electrophysiological conformation is, of course, found only in rhythmically active visceral muscles. In normal circumstances, the spike-like part of the action-potential response is found at the apex of the slow wave (see figure 4.1), i.e. at the point of maximum depolarization; this suggests that the slow wave is some form of pre-potential or generator potential which acts to move the membrane potential to the spike-firing threshold, at which point a spike is generated by regenerative ionic entry into the membrane. It is usually only after osmotic stress, potassium depolarization, or drug treatments, that the spike is displaced from the wave crest, but in these conditions the membrane ionic conductances are probably far from the normal resting values. It is possible, although rather unlikely, that the spike and slow wave may have arisen in different cells, the spike spreading electronically to the recording point, since the length constant often exceeds the length of a single cell.

Visceral muscles with the slow-wave/spike-action potential response still vary in the number of spikes which may occur per individual wave. In many preparations, one slow wave generates one spike only at its crest, exemplified by the cat duodenum, while in other muscles, the number of spikes per wave may vary, both from animal to animal and at different times in the same tissue. The rat ileum is a good example of this (figure 4.1), where one to three spikes may occur per wave, and in the longitudinal muscle of the rabbit duodenum, anything from one to eight spikes may be present per wave (see figure 4.2). Although there is little relation between the duration of the slow wave and the number of spikes it generates, there is a strong correlation between the rate of slow-wave depolarization and the number of spikes present. It is clear from figure 4.1 that slowly-depolarizing slow waves tend to have fewer spikes present than fast-depolarizing waves. This correlation may also be a measure of the size of slow-wave depolarization, and the time during which the wave exceeds the spike firing threshold, since large slow waves consistently generate more spikes than small slow waves.

The advantage of repetitive spike firing is to prolong the active state of the cell (see Chapter 6) and thus to enhance its contractile output. However, it must be remembered that contractions will still occur, although at a rather reduced level, in the ileum with few or even totally aborted spikes (see figure 4.1). It is clear that the actual function of a muscle within the body will determine its necessary contractile output, and thus its spike firing frequency and slow-wave characteristics.

A further modification of action-potential shape seen in spike-producing muscles is exemplified by the cockroach rectum (Nagai and Brown,

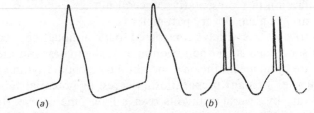

(a)   (b)

Figure 4.5. Differences in slow wave shape in single-unit visceral muscles.
(a). Slow 'pacemaker' wave with a spike which returns the membrane potential to maximum polarity. Typical of cockroach rectum and rat mesenteric artery.
(b). Sinusoidal-type slow wave with spikes superimposed at the crest. The spikes do not return the membrane potential to maximum polarity. This is typical of vertebrate gut muscles.

Table 4.3.—The magnitude and duration of the slow wave and spike

Animal and muscle	Slow wave (mV)	Wave duration (msec)
rat mesenteric blood vessels	3–9	300
rat ileum	10–14	2,500
cockroach rectum	6–8	2,000
carp stomach muscle	10–15	5,000
guinea-pig taenia coli	4–8	1,000
rat uterus	4–10	500
cat duodenum	8	4,000

1969) and rat mesenteric artery (Steedman, 1966). In these preparations, the slow wave takes the form of a simple spontaneous decay of the membrane potential, and this generates a single spike when the membrane potential reaches the spike firing threshold (see figure 4.2). Here, the slow wave takes the form of a pacemaker pre-potential typical of that seen in cardiac muscle and quite unlike the sinusoidal slow waves of most spontaneously active intestinal muscles (figure 4.5). In the sinusoidal type of slow wave (figure 4.5b), the spike is simply superimposed on the wave, and the spike does not return the membrane potential to resting value, this being achieved by the slow wave itself. In the generator-like slow waves, the spike falling phase returns the membrane potential to resting value, and hence the slow wave possesses only a depolarizing action. This type of action potential conformation is usually called the 'relaxation oscillator type'. It is uncommon, and not enough work has been done on this type of preparation to allow us to advance any hypothesis about the relationship between the spike and the slow wave. It would seem that the spike and slow wave are far more interdependent in this type of muscle than is seen in the majority of muscles. Table 4.3 summarizes what little is known about the relationship between the slow wave, spike size and duration in representative visceral muscles, and it can be seen just how variable almost every parameter is.

The electrical events we have considered so far, although they constitute the major spontaneous and non-spontaneous fluctuations in membrane potential, are by no means the only membrane potential rhythms in many visceral muscles. In many preparations, the resting potential value of the cells may vary by several millivolts over a long time course, this being particularly seen in genito-urinal and intestinal muscles. These long-term fluctuations in membrane potential may influence the rate or actual incidence of spike-firing slow waves, and it is probable that these long-term membrane-potential fluctuation cycles are correlated with cyclical in some visceral muscles.

Spike (mV)	Spike duration (msec)	Reference
15–20	50	Steedman (1966)
10–20	50	Syson and Huddart (1973)
30	500	Nagai and Brown (1969)
15–20	200	Ito and Kuriyama (1971b)
40–50	100	Casteels and Kuriyama (1966)
50	50	Marshall and Miller (1964)
6–10	100	Kobayashi et al. (1967)

changes in the general tonus of the organ to which the cells belong. Hormonal changes are certainly known to regularize long-term membrane-potential fluctuations in the uterus, and this may also be the case in other muscles. It is probable that these cyclical fluctuations in membrane potential level, which in turn cyclically raise and lower the excitability of so many visceral muscles, are related to gross physiological cycles in the body as a whole, mediated by blood-borne agents and controlled by little-understood circadian physiological rhythms. This complication alone adds a further dimension to the physiological variability of visceral muscles.

Slow-wave/spike interrelationship

The relationship between the spike and slow wave is rather confused. However, with the exception of the few known cases of generator slow waves, it is clear that, when spikes occur, they do not appear to change either the frequency or the amplitude of the slow waves, and this is suggestive of quite separate electrogenic bases for these electrical events.

That the slow waves and the spikes are separate electrogenic events is also indicated by the action of chemical agents which can modify or block the spike, leaving the slow wave virtually unaffected (e.g. caffeine, noradrenaline, neostigmine and acetylcholine). Slow waves even continue in the presence of agents which both block spikes and inhibit contraction, confirming the hypothesis that slow waves act to change cyclically the excitability of the fibre membrane. This permits a further but quite separate electrogenic process to occur, which develops a faster but more labile depolarization. In many visceral muscles it is this labile depolarization which engages the excitation-contraction coupling process, leading to strong tension generation.

The physiological basis of slow-wave generation is not understood. It is unlikely that these waves are neurogenic in origin, since they exist in totally anervous tissue such as chick amnion and many arterioles and veins. Furthermore, their frequency and amplitude are not usually affected by agents which excite or inhibit nerve-mediated activity. It is also unlikely that stretch is an agent in slow-wave activity, since waves will continue in isolated flaccid intestinal strips, and although stretch will trigger spike bursts in normal visceral muscles, it rarely affects the frequency or amplitude of slow waves.

It would seem far more likely that the basis of slow-wave generation lies in some electrogenic process inherent in the muscle-fibre membrane itself. It is quite possible that this process may be similar to pacemaker potential electrogenesis in cardiac muscle, and this may be why slow waves are relatively independent of drugs and membrane potential level. Slow-wave frequency may simply be a measure of the instability of the cell's membrane potential, this being characteristic of a particular preparation in a particular organism. If this is the case, and slow waves are due to cyclical membrane-potential instability, this may mirror the cell's metabolism and could be effected via cyclical acceleration and inactivation of the membrane sodium/potassium exchange pump. Inactivation of the pump will cause a sodium influx into the cell, resulting in depolarization, while acceleration of the pump will cause a potassium-enriched cell resulting in hyperpolarization. These changes will be rapid if the exchange pump is electrogenic. However, no data are available on membrane conductance changes during the slow waves, so to implicate an increased sodium conductance as the basis of slow-wave generation is somewhat speculative at this time. Nor are any data available on membrane Na/K exchange during the slow wave. A critical test of the hypothesis linking slow-wave production to cellular metabolism, particularly to the sodium pump, is the effect of metabolic inhibitors. Both spikes and slow waves are abolished in rat ileum by 10^{-4}M 2:4 dinitrophenol, which does suggest a metabolic origin of the membrane-potential instability responsible for slow-wave generation. However, this finding will have to be confirmed in other visceral muscles before it can be considered to be of general application.

Conduction of the action potential

So far we have considered only the basic electrical phenomena of the fibre membrane. Most visceral muscles consist of short fibres running in many planes, and electrical conduction takes place through the muscles as if they were syncytia. Action potentials spread by local circuit current through the well-known low-resistance tight junctions or nexuses (see Chapter 1), the latter acting as electrical couplers between the units of what is in reality a multi-unit electrical complex. To understand the nature and conduction characteristics of the visceral-muscle action potential, reference has first to be made to the physical passive electrical characteristics of the fibre membrane, and to how the membrane itself can be considered as an electrical circuit.

Membrane circuitry and cable conduction

Not only is the muscle fibre membrane electrically polarized, its electrical potential difference is maintained within fairly constant limits, despite the fact that ionic movements, and hence electrical charges, continually take place across the membrane. Since the membrane consists of outer and inner protein layers separated by an inner lipid core, it possesses two good electrical conducting media separated by a poor conductor. Membranes thus have the properties of a capacitor, and they are able to store charge and develop a potential difference across their 'plates'. The plates of this natural capacitor are charged by the diffusible ions which move across it according to their electrochemical gradients; since lipids have high dielectric constants, membranes tend to have high capacitance values, and can store surprisingly high charges. Membranes also show properties of resistance, and in its electrical behaviour, the muscle-cell membrane can be likened to a capacitor with a resistor in parallel. Since the resistor will allow both charging and discharging of the capacitor, changes in membrane resistance allow voltage changes to take place across the membrane. Since the resistor element of the membrane will allow some current to flow across it, it is a constant drain on membrane charge. That the membrane remains charged for long periods at a fairly constant level, even though it is analogous to a leaky capacitor, indicates that some continuous charging process must be in operation—in other words, a membrane battery must be present. We can summarize these simple electrical properties of the membrane in terms of a circuit diagram (figure 4.6a). Since the battery is an ion battery which charges the membrane by ionic movement, all electrogenically active ions should be represented as separate elements in this battery. Furthermore, the resistor is also a multiple unit, since the membrane possesses differential resistances to the passage of different ion species. The membrane will thus contain ion-species-specific resistor elements arranged in parallel

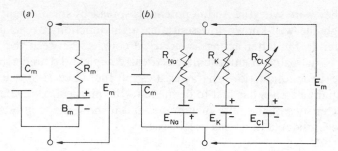

Figure 4.6(a). Elementary conceptual circuit diagram of a muscle-fibre membrane. The potential difference (E_m) across the membrane capacitor (C_m) is maintained by the membrane battery (B_m). The capacitor leaks via the membrane resistance (R_m).

(b). Expansion of this diagram to include serially arranged battery and resistor elements for the three common electrogenic ions arranged in parallel with the capacitance. The value of the capacitance charge (E_m) will vary as R_{Na}, R_K and R_{Cl} vary.

and ion-species-specific batteries also in parallel. Figure 4.6b shows a more sophisticated circuit model of the membrane but, since capacitance is not ion-specific, it is still represented as a single entity in parallel with the battery and resistor elements. Since the resistance to ions shown by the membrane is variable on a temporal basis, ion-specific resistors are represented as variable potentiometers.

The specific ionic resistance of the membrane is simply a measure of the permeability of the membrane to that ion. Membrane ionic permeability is usually expressed as *specific conductance*, although these terms are not exactly equivalent, and, since this is the ease with which the ion passes through the membrane, it is the reciprocal of resistance. The muscle-fibre membrane possesses different conductances to different ions at any one time, and the conductance of a specific ion may also vary with time. The specific membrane conductance or permeability to an ion is highest at its equilibrium potential, when the electrochemical driving force is zero. Thus, if a membrane is, for example, substantially depolarized away from the potassium equilibrium potential, then the driving force on the potassium ion will be high; but this will wane as the potassium permeability increases and the membrane potential approaches the potassium electrode potential. For the potassium ion, the membrane potassium conductance can be expressed as:

$$g_K = \frac{I_K}{E - E_K}$$

where I_K is the membrane potassium current, E is the membrane potential and E_K the potassium equilibrium potential. It can be seen that the lower the potassium conductance, the greater is the driving force $(E - E_K)$. This assumes that I_K remains fairly constant. Generally, when g_K falls I_K also falls, but this invariably leads to depolarization. Sodium and chloride conductances are represented in a similar fashion:

$$g_{Na} = \frac{I_{Na}}{E - E_{Na}} \quad \text{and} \quad g_{Cl} = \frac{I_{Cl}}{E - E_{Cl}}$$

where I_{Na} and I_{Cl} are the sodium and chloride membrane currents respectively, and E_{Na} and E_{Cl} are the sodium and chloride equilibrium potentials respectively. The conductance of the muscle-fibre membrane to these various ions shows differential variability, and in skeletal muscle, two common conditions pertain, the resting stable condition dominated by the potassium conductance, and the transient active membrane, dominated by the sodium conductance. The electrogenic mechanisms of visceral muscle are somewhat different from those of most skeletal muscles, and these will be considered later in this chapter.

Cable conduction

The way in which an electrical event spreads in the muscle-fibre membrane is dependent upon the electrical properties of the membrane. If part of the membrane becomes depolarized by inward ionic flow, a lateral redistribution of charge occurs, and current flows away from the depolarized point, depolarizing nearby regions of the membrane and activating more ionic carriers in the membrane. If these nearby regions become sufficiently depolarized, ionic inrushing occurs, and these regions in turn activate the next region of the membrane by local current flow.

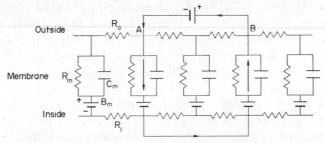

Figure 4.7. Simplified presentation of the muscle-fibre membrane as a series of resistor/capacitor/battery segmental units to explain cable conduction. An applied potential, or natural action potential, at A induces current flow (depolarization) at B. R_i and R_o are the internal and external fluid resistivities. R_m, C_m and B_m have their usual notation.

Hence, if part of the membrane is sufficiently depolarized, it creates an ion sink, causing outward current flow at some distant point, and resulting in critical depolarization and action-potential production at that distant point. This method of action-potential spread, depending upon lateral current flow from one segment of the membrane to another, is called *cable conduction*. Action-potential spread can best be visualized by considering the membrane to consist of a series of battery/resistor/ capacitor units, summarized in figure 4.7. Current flow involves the transverse resistance of the membrane, as well as internal (myoplasmic) and external (plasma or saline) resistivities, and the main concept of cable conduction is that one segment of the membrane depolarizes a distant segment, and short-circuit current flows between them.

Rushton (1937) showed that if the cell is assumed to be a cylinder, it is possible to calculate the change in potential (P) caused by an applied potential such as the action potential (P_o) at a distance x from P_o. He derived the relation:

$$P = P_o \, e^{-x/\lambda}$$

where λ, the length constant, is given by

$$\lambda = \sqrt{\frac{R_m}{R_i + R_o}}$$

where R_m is the transverse resistance of the membrane, and R_i and R_o are the internal and external fluid resistances. As shown in figure 4.7, a potential P_o at A causes a change in potential (P) at B, due to short-circuit current flow, hence an action potential depolarizes the membrane a certain distance ahead of it. The length constant simply determines how far B can be from A, while still allowing critical depolarization and action-potential regeneration. It is thus the value of the length constant of a muscle cell which determines the distance over which an action potential can critically depolarize the membrane to regenerate the action potential. Action-potential conduction velocity will thus depend upon the resistance properties of the membrane. The larger the length constant, the faster the action potential will spread; and since λ increases with cell diameter, the larger the fibre, the faster will the action potential spread. Small muscle fibres with high membrane resistances have small values for λ, and thus have slow conduction velocity. Most visceral-muscle fibres fall into this category. In many vertebrate visceral muscles, the length constant is greater than the length of a single cell (possibly caused by the presence of nexuses), and this makes the tissue an electrical syncytium since it greatly facilitates current spread.

The electrical characteristics of the visceral-muscle fibre membrane

Table 4.4 summarizes membrane transverse resistance and capacitance values in a selection of visceral muscles, and it can be seen just how variable these values are. Measurement of the longitudinal internal resistance of visceral muscle has yielded very variable results, due to pitfalls in technique and variations in methods of approach (see Tomita, 1970). However, the longitudinal resistance does seem to be low; in mammalian taenia coli a value of 400 ohms/cm has been reported.

This figure includes the myoplasmic resistance (100–200 ohms/cm) and the cell junctional resistance (about 300 ohms/cm). Longitudinal resistance values as low as this are to a large extent responsible for the great facility with which action-potential conduction occurs both along and between cells.

The membrane capacitance of visceral muscle is rather low compared to that of skeletal muscle. The difference in capacitance is almost certainly related to the T-system, which is well developed in skeletal muscle and usually completely absent in visceral muscle (with the exception of arthropod gut muscles). Capacitance is itself related to total muscle cell surface, and the T-system constitutes a considerable fraction of the latter. The low capacitance of visceral muscle is simply a measure of the poor ability of the membrane to store charge, this being correlated with the generally low resting-potential values.

Earlier in this chapter it was stated that in muscle-fibre membranes the action-potential spread could be explained in terms of simple cable theory. To close this section we can now examine some of the evidence for this view in relation to the visceral-muscle fibre. In frog stomach muscle, guinea-pig taenia coli, vas deferens and portal vein, induced electrotonic potentials spread with almost perfect exponential decay, the potential spreading in a manner related to the length constant of the fibres, and falling dramatically with distance from the point of initiation. Data on this point can be seen in figure 4.8. This kind of spread of membrane depolarization fits almost perfectly the concept of 'segmental' membranes composed of capacitor/resistor/battery units. Furthermore, if the membrane is a cable-like structure, the initial depolarization of the spike should rise exponentially, the time-constant being determined by the cable properties of the membrane. That this is in fact the case has

Table 4.4.—Major membrane electrical characteristic of some

Animal and muscle	Specific membrane resistance (ohm/cm²)
cat intestinal muscle	1,050
cat intestinal longitudinal muscle	780
guinea-pig taenia coli	320
guinea-pig jejunum and rectum	600–650
guinea-pig ureter	300
cockroach rectum	500 k–5 M

* Length constants (in mm) reported are cockroach rectum (1·5 to 4); guinea-pig

been most elegantly demonstrated for the guinea-pig taenia coli action potential by Tomita (1966). The final confirmatory evidence of the application of cable theory to visceral-muscle membranes is provided by the results of voltage clamp studies. In voltage clamping, the membrane is maintained at a 'holding potential' via current fed through an intracellular electrode. With a second electrode, depolarizing and hyperpolarizing pulses can be applied to the membrane, and the behaviour of the induced ionic currents studied. An important voltage-clamp study of rat uterine muscle by Anderson (1969) has revealed much about membrane ionic-conductance mechanisms in visceral muscle. These

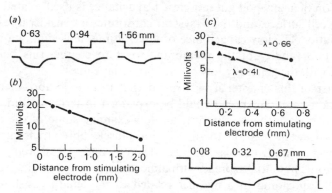

Figure 4.8(*a*). Electrotonic potentials (lower trace) produced in guinea-pig taenia coli by externally applied hyperpolarizing currents. The potentials were recorded at three different distances from the stimulating electrode (redrawn after Tomita, 1970).
(*b*). Spatial decay of the electrotonic potential of the taenia coli (from data in Abe and Tomita, 1968).
(*c*). Spatial decay of the electrotonic potential of fibres of the guinea-pig portal vein, drawn at the maximum and minimum values of the length constant. The traces below the graph show electrotonic potentials at three different distances from stimulus point, with a length constant of 0·66 mm. Redrawn from data in Ito and Kuriyama (1971*a*).

visceral muscles.*

Membrane capacity (μF/cm²)	Reference
21	Kobayashi *et al.* (1967)
13	Kobayashi *et al.* (1967)
10	Kuriyama and Tomita (1965)
4·5–5·0	Kuriyama *et al.* (1967*b*)
8	Kuriyama *et al.* (1967*a*)
0·4–3·5	Belton and Brown (1969)

taenia coli (1·5); guinea-pig vas deferens (2·0).

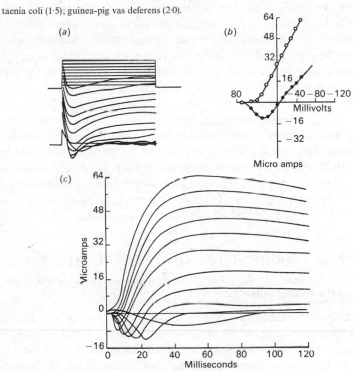

Figure 4.9. Voltage clamp studies of rat uterine muscle.
(*a*). Series of membrane currents in response to step-wise changes of membrane potential from the holding potential of −70 mV to +20 mV.
(*b*). Plot of current/voltage relations of the peak transient current (filled circles) and steady-state current (open circles) after correction for leakage current.
(*c*). Traces of membrane currents (after correction for leakage current) under voltage clamp conditions. At +30 mV, the transient current begins horizontally, indicating that this is about the value of the transient-current equilibrium potential. The membrane potential in these traces reads from bottom to top in 10 mV stages from −50 mV to +50 mV. All redrawn from data in Anderson (1969).

Table 4.5.—The potassium content, potassium electrode potential

Animal and muscle	K(mM/kg tissue water)	K_o(mM)
guinea-pig taenia coli	86·4	5·9
cat intestine	108	5
rat uterus	143	3
guinea-pig ureter	154	4

experiments have shown that the membrane of the uterine cells develops a series of ionic currents with voltage-dependent and time-dependent characteristics as the membrane potential is progressively displaced away from the holding potential. The rate and mode in which the family of ionic currents decay in relation to the displacement of the membrane potential confirms the view that the passive electrical properties of visceral-muscle fibres can be expressed in terms of cable theory. Figure 4.9 illustrates ionic current recordings from the rat uterine muscle-fibre membrane. After an initial capacitative leakage current, a sharp transient inward current, and then a delayed steady-state outward current is seen, these currents altering as the membrane potential is displaced from the holding potential. It can be seen that the current/voltage relation is a smooth and graded continuous function of membrane potential, a condition which would be expected on cable-theory postulates.

THE ELECTROCHEMISTRY OF VISCERAL MUSCLE

Ionic basis of the resting potential

As in other muscle tissues, the resting-membrane potential of visceral muscle is explicable in terms of the differential distribution of the common ions across the cell membrane and the relative permeability of the cell membrane to these ions. According to the ionic hypothesis, the resting potential of skeletal muscle is largely determined by the passive distribution of potassium, the potassium resting conductance being high, so that the membrane behaves as a potassium electrode. In these conditions, the resting-membrane potential will be approximately equal to the potassium electrode potential, given by the Nernst equation:

$$E_m = E_K = \frac{RT}{zF}\log_{10}\frac{(K)_i}{(K)_o}$$

and resting potential of some visceral muscles.

E_K(mV)	Resting potential (mV)	Reference
−92	−57	Casteels and Kuriyama (1966)
−80	−52	Barr (1959) Burnstock and Prosser (1960b)
−87·4	−40	Cole (1950) Anderson et al (1971)
−98	−59	Washizu (1969)

At 20 °C, this equation becomes:

$$E_K = 58\,\text{mV}\,\log_{10}\frac{(K)_i}{(K)_o}$$

where R is the gas constant, T the absolute temperature, z the valency of the ion and F the Faraday constant. The passive resting distribution of potassium also seems to be an important factor in resting-potential electrogenesis in visceral muscle. As will be seen from the Nernst equation, an increase in external potassium should lead to a depolarization of the membrane, the depolarization slope being about 58 mV for a tenfold potassium increase. The effect of external potassium on the resting potential of a variety of visceral muscles can be seen in figure 4.10, and it can be seen that K-induced depolarization is far less than would be expected if the membrane behaved as a good potassium electrode. In

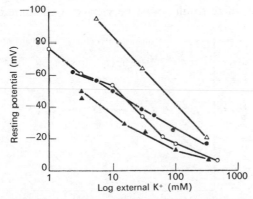

Figure 4.10. The effect of external potassium on the resting potential of some visceral muscles. Closed triangles, rat ileum (Syson, previously unpublished); open circles, guinea-pig ureter (Washizu, 1966, 1967); closed circles, guinea-pig taenia coli; open triangles, the calculated potassium electrode potential of the guinea-pig taenia coli (Casteels and Kuriyama, 1966).

preparations as diverse as rat ileum, guinea-pig taenia coli, vascular muscle and insect rectum, the depolarization slope for a decade potassium change is only in the 30–35 mV region.

Ionic analysis of visceral-muscle myoplasm enables the true K electrode potential (E_K) to be calculated from the measured values of myoplasmic (K_i) and plasma (K_o) potassium levels. Data bearing on this point are shown for some selected visceral muscles in Table 4.5, along with recorded values of the resting potential. It can be seen that in no preparation does the E_K approach the actual membrane potential, and this could be due to a variety of reasons. Firstly, there is evidence that myoplasmic potassium changes as the external potassium is altered in a manner not predictable by the Nernst equation; and secondly, it is now clear that visceral muscle has a far higher resting chloride conductance than skeletal muscle, and chloride ions appear to make some significant contribution to the resting potential (Casteels and Kuriyama, 1966). It is difficult to assess exactly what contribution chloride may make, since K_i/K_o and Cl_o/Cl_i reciprocity, as would be expected from Nernst postulates if the membrane were freely permeable to K and Cl and E_K equalled E_{Cl}, does not apply in these muscles, but chloride replacement has a significant effect on resting potential.

A third complication in analyzing the electrogenic mechanism behind the resting potential is caused by the presence of a large resting sodium conductance in visceral muscle, suggesting that sodium ions may contribute to the resting potential. Sodium replacement causes a depolarization of the resting potential in taenia coli and guinea-pig vas deferens (figure 4.11); but these results must be viewed with some caution, since

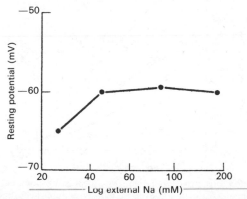

Figure 4.11. The relationship between external sodium and the size of the resting potential in guinea-pig vas deferens. These low sodium solutions were not adjusted for lowered chloride or osmolarity. From data in Bennett (1967).

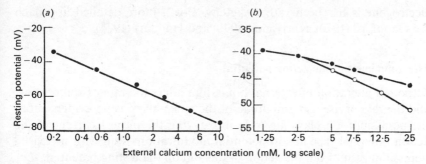

Figure 4.12. The relationship between external calcium concentration and the resting potential of guinea-pig vas deferens (a) and cat ureter (b). In cat ureter, resting potential changes were recorded in normal sodium (open circles) and sodium-free (closed circles) salines. Redrawn from data in Bennett (1967) and Kobayashi (1969).

alteration of sodium may affect the conductances of other ions such as calcium (Kuriyama, 1970), which may in turn make some electrogenic contribution to the resting potential. In guinea-pig taenia coli, and vas deferens, and in the cat ureter, calcium removal depolarizes the membrane, and a slope of about 25 mV is seen for a tenfold external calcium change (figure 4.12).

The obvious involvement of several ionic species in the maintenance of the resting potential of visceral muscle makes it likely that the Goldman constant-field equation would give a better prediction of the resting potential than the E_K (which is too high) or the E_{Na} and E_{Cl} (which are too low). According to this view, the membrane potential can be expressed as:

$$E_m = \frac{RT}{zF} \log_{10} \frac{P_K(K)_i + P_{Na}(Na)_i + P_{Cl}(Cl)_o}{P_K(K)_o + P_{Na}(Na)_o + P_{Cl}(Cl)_i}$$

where P represents the membrane relative permeability to the ion involved, and the other terms have their usual notation. Casteels (1969) examined the application of this equation to the guinea-pig taenia coli resting potential, using relative permeability ratios of $P_K:P_{Cl}:P_{Na}$ of 6·1:3·7:1·0 respectively. The result, at −37 mV, is about 20 mV below the recorded potential, and Casteels (1969) proposed that the resting potential may be only partly due to passive ion distribution, and partly due to an electrogenic sodium pump, an active transport mechanism dependent upon cellular metabolism. This does seem a probable explanation, and it has already been proposed to explain the anomalous resting potentials of some insect muscles. Of necessity, the account given here of resting potential electrogenesis has been rather sketchy. For a more detailed approach see Goodford (1970), Kuriyama (1970) and Huddart (1974).

The ionic basis of the action potential

Although alteration of external potassium alters the rate of spike firing, and the rate of rise and amplitude of the spike, there is no evidence that the potassium ion plays any role in carrying the inward current of the action potential. The action-potential modifications seen with increased potassium result from the action of this ion on the resting potential.

According to the ionic hypothesis, the action potential of skeletal muscle is due to a specific increase in sodium conductance, so that at its peak the membrane behaves as a sodium electrode; the action-potential overshoot is explicable by the sodium version of the Nernst equation:

$$E_m = E_{Na} = \frac{RT}{zF} \log_{10} \frac{(Na)_o}{(Na)_i} = 58\,mV \log_{10} \frac{(Na)_o}{(Na)_i} at\ 20\,°C$$

In visceral muscle, the critical test of the applicability of the ionic hypothesis is to alter the external sodium concentration and observe the effect on the action potential. In the majority of visceral muscles so far studied, the action potential is rather insensitive to even large changes in Na_o, the slope of action-potential height for a decade Na_o concentration change often being under 10 mV. This action-potential sodium insensitivity has been found in preparations as diverse as uterus, taenia coli and vas deferens. Figure 4.13 shows typical results from guinea-pig vas deferens; it can be seen that even large changes in external sodium were

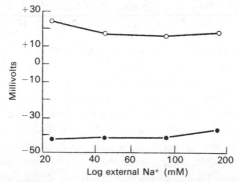

Figure 4.13. The relationship between external sodium and the size of the action-potential overshoot (open circles) and the threshold for the initiation of the action potential (filled circles) in guinea-pig vas deferens. Redrawn from data in Bennett (1967).

without effect on either the action-potential overshoot or the threshold
for spike firing. Confirmatory evidence for the non-dependence of most
visceral-muscle action potentials on sodium is provided by the general
insensitivity of the action potential to tetrodotoxin. This latter toxin acts
specifically to block the membrane channel involved in sodium tran'port
into the muscle fibre, and it has been used as a bioassay for sodium-
dependent electrogenesis. In visceral muscle as a whole there are few
positively documented cases of sodium-dependent action-potential
electrogenesis; some other electrogenic mechanism must be postulated.

There is now overwhelming evidence that calcium ions are the main
agents involved in inward current carrying for spike generation in most
visceral muscles. In sodium-free saline, spontaneous action potentials can
be restored by addition of excess calcium in ureter, taenia coli and
stomach muscle. In most visceral muscles, lowering the saline calcium
both reduces spike amplitude and rate of rise and increases cell-membrane
resistance and the threshold for the initiation of the action potential
(figure 4.14). In the cat intestine, a comprehensive study by Job (1969)
conclusively showed that calcium ions were the principal current carriers
for the depolarization phase of the spike potential.

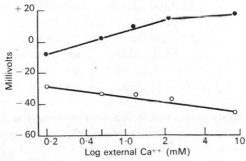

Figure 4.14. The relationship between the size of the action-potential overshoot (filled circles)
and the threshold for the initiation of the action potential (open circles) and external calcium
concentration in guinea-pig vas deferens. Redrawn from data in Bennett (1967).

In the case of the slow wave, a quite separate electrogenic mechanism
appears to be involved. In the taenia coli, the slow wave is suppressed
in sodium-free saline, while in sodium-rich salines the slow wave is
increased in size, and repetitive spikes are often seen (Bulbring and
Kuriyama, 1963). In the duodenum, and the cat and guinea-pig stomach,
the slow wave is significantly reduced if saline sodium is reduced by as
little as only 20%, and is blocked completely in sodium-free saline.

In a more critical study, Job (1969) has shown that the slow wave of cat intestine is correlated with a passive sodium influx into the cells. It would seem that the slow wave originates in a cyclical variation in the activity of the membrane sodium pump, permitting depolarization to a level which triggers increase in calcium conductance leading to the spike.

It must be borne in mind that in many muscles of the genital tract, both resting and action potentials may be modified by the hormonal state of the animal. In the non-pregnant rat and cat uterus, the spike rarely overshoots zero potential, but in the pregnant state, a considerable overshoot is seen. In the case of the resting potential, the balance between progesterone and pituitary oxytocin determines the degree of hyper- or depolarization, and hence the excitability of the cells. In such tissues, although the slow waves and spikes are sodium- and calcium-dependent respectively, steroid hormones appear to modify the muscle-fibre membrane conductances to these ions.

The ionic mechanism underlying the repolarization phase of the visceral-muscle action potential has received little attention. In skeletal muscle, repolarization is brought about by a sharp increase in potassium conductance, and a potassium efflux from the fibre. This has been confirmed by the use of TEA (tetraethylammonium) ions, which specifically block the membrane potassium channels, reducing the potassium conductance. Application of TEA ions to skeletal muscle converts normal action potentials into plateau-like action potentials due to retarded repolarization (anomalous rectification). A similar situation has recently been found in guinea-pig stomach muscle and rabbit common carotid arteries, where TEA ions suppressed the rectifying properties of the membrane. There would seem to be little doubt that in these preparations at least, and probably in most other visceral muscles, the action-potential repolarization phase is due to an increase in the potassium conductance of the membrane.

CHAPTER 5

EXCITATION-CONTRACTION COUPLING

Excitation-contraction (EC) coupling is the name given to a whole series of processes, both electrical and chemical, which link the transient depolarization of the muscle-fibre plasma membrane to the activation of the myofilaments and the development of tension. The relaxation part of the muscle-fibre cycle can be simply considered as part of the EC-coupling process in reverse. It was clear even to the earliest investigators that *something* must link depolarization to contraction, but exactly what was involved remained something of a mystery until quite recently. It had been a *consistent* observation with skeletal, cardiac and visceral muscles, confirmed after the widespread application of intracellular microelectrode techniques, that contraction of a muscle fibre would occur only after the cell was first activated by a depolarizing electrical disturbance of the plasma membrane—whether this be caused by a naturally-occurring action potential or graded response, by artificially-applied stimulation, or by increased KCl levels. Certain rigour promoting drugs such as caffeine, quinine, nicotine, organophosphorous and organochlorine compounds are able to induce contraction without transient depolarization, but these are special cases which will be considered later. Clearly depolarization triggers contraction, but how the trigger works is not at all obvious. The major problem which confronted the early investigators was that contraction followed depolarization so very rapidly it was evident that any translocation of some sort of activator agent from the plasma membrane itself to the interior of the fibre would be far too slow to explain the speed of onset of contraction, or myofilament activation at least, even in the small fibres of visceral muscles. Almost all work on EC-coupling has been carried out on skeletal muscle, and a little on cardiac muscle, so what is known about EC-coupling in these cases will be reviewed first, leading on to the specialized case of visceral muscle.

That the processes involved in EC-coupling are not simple, but a complex mixture of electrical and chemical events involving several separate physical parts of the muscle cell, is shown by the many ways in which EC-coupling can be disrupted, and excitation uncoupled from contraction. Calcium deprivation, for example, while allowing almost normal-action potential electrogenesis, inhibits contraction at a later coupling stage. Selective disruption of the transverse tubular system by glycerol-induced osmotic shock also uncouples excitation from contraction (Howell, 1969), but without altering cellular calcium fluxes (Van Der Kloot, 1968). Structural modification of the sarcoplasmic reticulum by drugs such as caffeine, quinine, organophosphorous and organochlorine agents also modifies coupling (Huddart and Oates, 1970; Bradbury, 1973a, b). In these various treatments, quite separate stages of EC-coupling are modified or extinguished. This preliminary glance makes it clear that EC-coupling consists of a series of interlinked stages, and modern experimental investigation from a variety of unrelated sources, reviewed excellently by Sandow (1965, 1970) and by Bianchi (1969) and Huddart (1974), has established the nature and sequence of the major events in the coupling process in skeletal muscle. These can be briefly summarized as:

(1) Coupling begins with a transient triggering depolarization of the cell surface membrane beyond a critical threshold value (the *mechanical threshold*).

(2) This is followed by a transverse spread of electrical excitation into the interior of the muscle fibre.

(3) The internal depolarization activates the sarcoplasmic reticulum, leading to a release of some of its stored calcium.

(4) Calcium diffuses to the contractile apparatus, resulting in tension generation.

The relaxation cycle contains one further stage—the resequestering of calcium from the sarcoplasm by the sarcoplasmic reticulum, resulting in myofilamentary de-activation. This is not a passive process; it involves metabolic energy, so relaxation is, in fact, an active event.

In the rest of this chapter, the coupling sequence will be examined stage by stage, including relaxation, and the special case of visceral muscles will be considered at each stage.

The mechanical threshold

The mechanical threshold can be defined as the membrane potential level at which the process of EC-coupling begins, i.e. the level required to initiate activation of the contractile elements. That this threshold level really exists can be shown by altering the muscle-cell resting potential and measuring tension output. This is best carried out on single fibres, to reduce scatter of results, and such an experiment can be seen in figure 5.1(a), from the data of Heistracher and Hunt (1969). Here, the membrane potential was subjected to step-wise changes using voltage clamp techniques, involving passing a current into the fibre through a microelectrode. It can be seen that little tension is generated at first, but depolarizations beyond $-40\,mV$ induce a steep generation of tension, maximum tension being generated

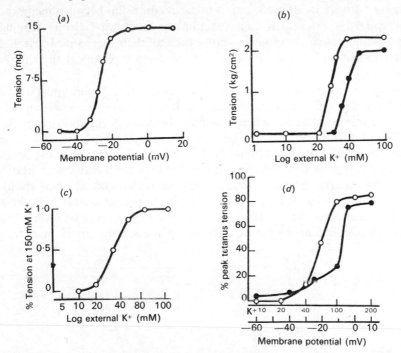

Figure 5.1. The mechanical threshold of skeletal muscle fibres.

(a) The relationship between tension and membrane potential in single fibres of snake skin. Membrane potential was altered by voltage clamping with a current-feeding microelectrode. Redrawn from data in Heistracher and Hunt (1969).

(b) The relationship between external potassium concentration and tension in frog semi-tendinosus fibres (open circles) (redrawn from data in Hodgkin and Horowicz, 1960) and locust spiracular muscle (closed circles) (from data in Hoyle, 1961).

(c) External potassium/tension plot in pigeon iris muscle (from data in Pilar and Vaughan, 1969).

(d) Tension/external potassium plots for locust muscle fibres (open circles) from data in Huddart and Abram, 1969) and crab muscle fibres (from data in Huddart, 1969). In all cases, a sharp mechanical threshold can be seen.

over a small depolarization range (about 15–20 mV). The membrane potential level at which the sharp increase in tension occurs is the mechanical threshold, which for this particular preparation (snake skin fibres) is at about −25 mV.

Most muscles have fibres too small for isolated fibre experiments, and thus for technical reasons, most studies on mechanical threshold have employed whole muscles, in which the membrane potential was altered by modifying the KCl content of the bathing medium. A selection of typical results from skeletal muscles of vertebrates and invertebrates is shown in figure 5.1(b–d). In all cases, a non-linear relationship between membrane potential (or external K-concentration) and tension is seen. Although individual muscles vary in the actual value of their mechanical-threshold depolarization level, in all cases, a clear sharp mechanical threshold is evident.

In a recent investigation of two visceral muscles, Syson and Huddart (1973) found that no sharp mechanical-threshold point was present. This can be seen in figure 5.2, where tension in both rat ileum and vas deferens is an almost linear function of increase in external potassium. Actual records of the tension responses can be seen in figure 5.3. These results may simply point to a different relationship between membrane excitation and the internal calcium-releasing agencies within the cell. Not enough visceral muscles have been examined so far to permit a clear hypothesis, but it does seem that in these muscles, the degree of activation of the contractile elements by the EC-coupling process is not an all-or-nothing

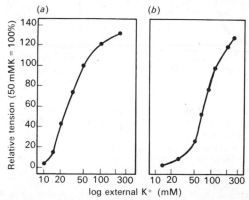

Figure 5.2. The relationship between external potassium concentration and tension in (a) rat ileal smooth muscle and (b) vas deferens smooth muscle. In both preparations, there is no evidence of a sharp discontinuity or mechanical threshold (from data in Syson and Huddart, 1973).

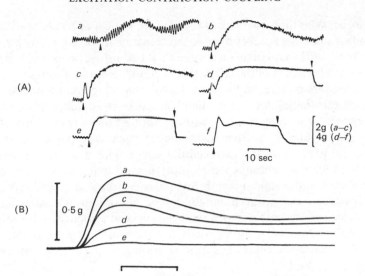

Figure 5.3A. Responses of rat ileal smooth muscle to potassium salines (*a*) 15 mM KCl; (*b*) 20 mM KCl; (*c*) 25 mM KCl; (*d*) 30 mM KCl; (*e*) 40 mM KCl; (*f*) 50 mM KCl. Calibrations 10 sec and 2g (*a–c*) or 4g (*d–f*).
B. Potassium-induced contracture tension in rat vas deferens smooth muscle. (*a*) 125 mM KCl; (*b*) 100 mM KCl; (*c*) 75 mM KCl; (*d*) 50 mM KCl; (*e*) 25 mM KCl. In both preparations, tension appears to be a continuous function of KCl-concentration. Both from Syson and Huddart (1973).

event, but is linear in relation to depolarization. In other words, a partial coupling can occur since no sharp mechanical threshold is present. In visceral muscles, a surface-mediated release of activating calcium seems highly probable due to lack of a highly organized T-system and sarcoplasmic reticulum (see p. 101). In these cases, a much more direct influence of membrane potential level on calcium release may be present, making an all-or-none trigger device unnecessary to switch on EC-coupling.

Depolarization spread in muscle cells

The mechanism of the inward transmission of surface excitation to the interior of muscle fibres became clearer when electron microscopists turned their attention to the fine structure of skeletal muscle fibres. These studies showed the presence of a complex set of transversely oriented tubules (the T-tubules or transverse tubular system) which ramified to the innermost parts of the fibres, making contact at regular intervals with another set of tubules—the longitudinally oriented sarcotubular system, more commonly called the *sarcoplasmic reticulum*. The specialized contact areas between these two sets of tubules are known as *triads* or *dyads*, depending upon the number of apposed elements involved. However, it was not until 1964 that Huxley was able to show a direct continuity between the central element of the triad (the T-tubule) and the plasma membrane in vertebrate muscle, and not until 1967 that Hagopian and Spiro confirmed this finding for one of the elements of the dyad in insect muscle. The timely studies on chloride withdrawal by Foulks *et al.* (1965) showed that the T-tubules, by their osmometric responses, possessed contents identical to those of the extracellular space. This evidence showed that the T-tubular membranes were simply inward extensions of the plasma membrane, with a similar membrane polarity, whose contents were simply extensions of the extracellular fluid. The physiological studies, coupled with electron-microscopic observation, was clear evidence to show that the T-tubules provided a direct route for the inward transmission of the transient surface-membrane depolarization into the interior of the fibre with great speed.

In the case of visceral muscle, there are considerable problems related to the spread of surface depolarization. With the exception of arthropod gut muscles, which conform to the above-mentioned structural pattern, no visceral-muscle fibres possess a well-organized T-system. Most visceral muscles are totally devoid of such a system, although a few invertebrate muscles (such as molluscan muscles) possess short tube-like extensions of the plasma membrane into the peripheral part of the sarcoplasm (see Chapter 1). It is evident that in most visceral muscles there is little dissemination of the transient surface depolarization into the fibre. The small vesicle or tube-like internal membrane projections certainly act to increase the membrane surface area over which current flows, but it is necessary to postulate a much more peripheral release of activating substance in the case of visceral muscle. As in skeletal muscles, visceral muscles are activated by a rise in myoplasmic free calcium, and it seems probable that the great bulk of this activating calcium has to be translocated from surface sites (either from the extracellular fluid through the plasma membrane or from storage sites immediately below the surface). It would therefore be expected that the activation process in skeletal muscle would be slow, and this is certainly the case compared with skeletal muscle (Syson, 1974).

Calcium storage and the state of calcium in muscle

As in skeletal muscle, it is a rise in myoplasmic free calcium which is responsible for the initiation and maintenance of contraction in visceral

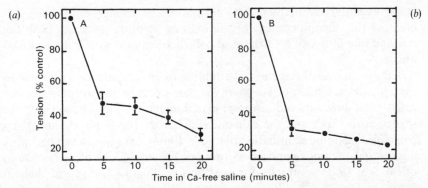

Figure 5.4. The effect of calcium-free saline on the size (*a*) and rate of rise (*b*) of 50 mM KCl contractures of ileal smooth muscle. Each point represents the mean ±S.E. (*n* = 5). Figure courtesy of Dr. A. J. Syson.

muscle. Whereas in skeletal muscle, the calcium rise comes largely from the calcium released from the sarcoplasmic reticulum at deep intracellular loci, in visceral muscles the extracellular calcium level is of great importance. The importance of extracellular calcium in mediating EC-coupling has been shown by Washizu (1967) and Bulbring and Tomita (1970), who incubated ureter and taenia coli muscles in calcium-free and calcium-EDTA media, and noticed a rapid abolition of mechanical responses. This is illustrated in figures 5.4 and 5.5 for rat ileum and vas deferens, and it can be seen that a continued supply of extracellular calcium

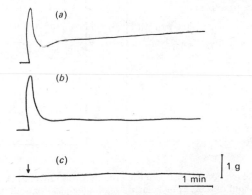

Figure 5.5. The effect of calcium-free saline on 125 mM KCl contractures of rat vas deferens smooth muscle. (*a*) Control contracture; (*b*) contracture after 30 sec in calcium-free saline; (*c*) contracture after 15 minutes in calcium-free saline. Calibrations 1 min and 1g. Figure courtesy of Dr. A. J. Syson.

is an essential requirement for normal EC-coupling. Conversely, increasing the calcium content of the bathing medium increases both the size and rate of rise of KCl-induced contractures of vas deferens and ileal muscle.

That the terminal stage of EC-coupling in visceral muscle is similar to that in skeletal muscle (i.e. involving an increase in myoplasmic free calcium) is shown by radioisotope experiments in which the muscles are loaded with Ca^{45}, and the calcium efflux from the fibres examined by conventional liquid scintillation methods. Figure 5.6 shows a typical efflux desaturation curve, expressed as rate coefficient (% loss of activity/min) of rat ileal smooth muscle. Visceral muscle differs a little from skeletal muscle as far as Ca^{45} efflux is concerned. Whereas skeletal muscle efflux curves invariably show the presence of a two-compartment system, visceral muscles have a three-compartment system. The two compartments of skeletal muscle correspond to efflux from the extracellular space (the fast initial component), the slow sustained component representing efflux from bound intracellular compartments. The clear three compart-

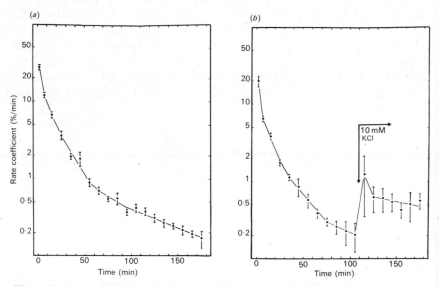

Figure 5.6. (*a*) The relationship between the instantaneous rate coefficient (%/min) and time (min) during washout of Ca^{45} from isolated longitudinal ileal smooth muscle. Each point is the mean $\pm$ S.E. ($n = 6$).

(*b*) The effect of 100 mM KCl saline substitution on the Ca^{45}. efflux rate coefficient of isolated rat ileal smooth muscle. Each point is the mean $\pm$ S.E. ($n = 6$). Both figures courtesy of Dr. A. J. Syson.

ments of visceral muscle reflect a real difference in the calcium-compart-mentalization system. It has been suggested that the fast system component emanates from calcium in the extracellular space, the medium compartment from plasma membrane sites, and the slow compartment from intracellular bound sources (Syson, 1974). If a loaded visceral muscle is allowed to lose calcium until the slow sustained efflux is established, representing loss from the intracellular compartment (after approximately 100 minutes), and then the effluxing medium is changed for one with a high potassium concentration, an immediate stimulation of calcium efflux is seen (figure 5.6b). This is clear evidence that the contractures seen due to KCl-depolarization are mediated via an increase in myoplasmic free calcium. Experiments of this type (see Syson, 1974) have shown that calcium utilization by the contractile machinery of visceral muscle is similar to that seen in skeletal muscle.

The major question as far as visceral muscle is concerned relates to the actual source of calcium needed for the activation of contraction, and how this calcium is removed, and exactly to where it is removed during relaxation.

In skeletal muscle, the most important breakthrough in our under-standing of EC-coupling occurred when biochemical studies were carried out on the sarcoplasmic reticulum. In the early 1960s it was shown that fragmented sarcoplasmic reticulum, isolated from muscle by homo-genization and differential centrifugation, retained much of its normal biochemical activity. In particular, isolated reticulum was shown to be able to relax (i.e. to lower the optical density of) isolated actomyosin, and this activity was shown to be related to a particular biochemical function, that of calcium sequestration. In other words, the sarcoplasmic reticulum possesses 'relaxing factor' activity. In a series of pioneering studies, Carvalho (1966, 1968) showed that reticulum possessed powerful calcium-pumping activity, exchangeably binding calcium for magnesium at reticu-lar Ca/Mg ATPase binding sites under various conditions. It quickly became clear that the sarcoplasmic reticulum provided a perfect physio-logical mechanism for cyclically raising and lowering myoplasmic free calcium—an activity already known to be a prerequisite for contraction and subsequent relaxation. The activity of the sarcoplasmic reticulum is thus regarded as the very heart of the EC-coupling mechanism. Figure 5.7 shows the appearance of typical isolated reticulum from skeletal muscle, negatively stained in phosphotungstic acid. Typical calcium-uptake curves for some skeletal-muscle reticulum preparations can be seen in figure 5.8, and it is now known that the calcium releasing and binding

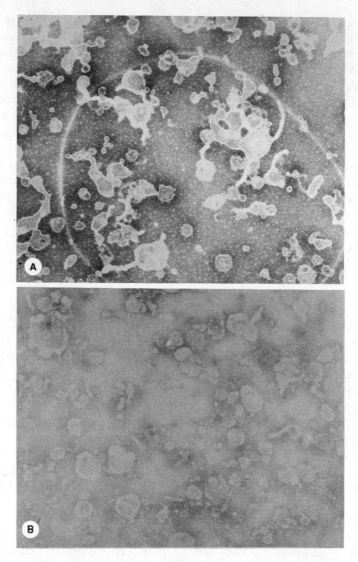

Figure 5.7. (a) Fragmented sarcoplasmic reticular vesicles isolated from locust skeletal muscle, negatively stained with 1% phosphotungstic acid and examined by conventional transmission electron microscopy. Print magnification × 34,000. Plate courtesy of Mr. M. Greenwood. (b) Plasma membrane vesicular fraction isolated from rat ileal smooth muscle, negatively stained with 1% phosphotungstic acid and examined by conventional transmission electron microscopy. Print magnification × 30,000. Plate courtesy of Dr. A. J. Syson.

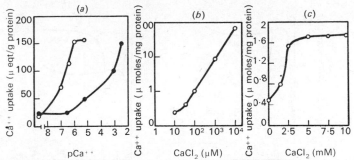

Figure 5.8. Calcium accumulation by sarcoplasmic reticular microsomes isolated from skeletal muscle.
(a) Rabbit skeletal muscle with ATP (open circles) and without ATP (closed circles). Redrawn from data in Carvalho and Leo (1967).
(b) Lobster skeletal muscle, redrawn from data in Van Der Kloot (1965).
(c) Ragworm skeletal muscle, redrawn from data in Stossel and Zebe (1968). In all cases, calcium uptake rises steeply with increase in calcium in the presence of ATP.

activity of sarcoplasmic reticulum is such as to account satisfactorily for the cyclical change of myoplasmic free calcium from 10^{-6} to 10^{-7} M, the levels responsible for maximal activation of contraction and full relaxation (see reviews by Sandow, 1965, 1970). As far as skeletal muscle was concerned, only two problems appeared to stand in the way of a full analysis of the entire coupling mechanism:

(1) How did the surface action potential induce the sarcoplasmic reticulum to release its bound calcium?
(2) How was the reticular calcium pump activated to resequester calcium in order to terminate contraction?

Evidence had already accumulated in the late 1960s that the T-system and the sarcoplasmic reticulum were closely connected, and evidence suggested that the action potential, or artificially applied depolarization, would rapidly spread to the extremities of the T-system. The discovery of junctional feet connecting the T-system with the lateral cisternae of the reticulum provided a mechanism (i.e. a low-resistance pathway) for the electronic spread of depolarization to the reticulum itself, and it is now known that depolarization of the sarcoplasmic reticulum induces a rapid release of its bound calcium in both amphibian (Endo and Nakajima, 1973) and mammalian (Miyamoto and Kasai, 1973) skeletal muscle fibres.

We are also much nearer to understanding how the reticulum may cyclically raise and lower myoplasmic calcium. Carvalho (1968a) had shown that calcium binding by the reticulum of rabbit muscle was influenced by the pH of the medium. However, Nakamaru and Schwartz (1972) showed that calcium exchangeability for magnesium at reticular

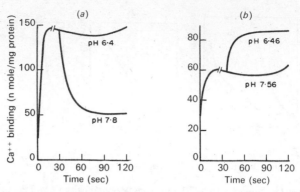

Figure 5.9. The effect of pH changes on calcium binding and release by sarcoplasmic reticulum isolated from dog skeletal muscle.
(a) Reduction in calcium accumulation with pH decrease.
(b) Increase in calcium accumulation with pH decrease. The break in the curves represents the point at which the pH was altered. Redrawn from data in Nakamaru and Schwartz (1972).

ATPase binding sites was exquisitely pH-sensitive over a very narrow range of only about one unit. Their results are summarized in figure 5.9. It can be seen that a pH-increase from 6·4 to 7·8 caused an abrupt release of bound calcium, while a pH-decrease from 7·5 to 6·5 caused a sharp increase in calcium binding. Carter *et al* (1967) had already demonstrated that muscle fibres transiently changed their pH as they were depolarized and then repolarized, suggesting that calcium activation of contraction and calcium removal for relaxation may be under the control of the action potential. The action-potential depolarization phase could be envisaged as increasing reticular pH, inducing calcium release, while the repolarization phase may lower pH, inducing calcium sequestration. Although this concept must be still regarded as tentative until further data are available, the overall model of EC-coupling control by the calcium modulation mechanism of the reticulum seems plausible enough.

The main problem in visualizing how a system such as this could work in visceral muscle is that most visceral muscles lack a well-developed sarcoplasmic reticulum. A form of sarcoplasmic reticulum has been claimed in some vertebrate smooth muscles (Somlyo, 1972). Studies with certain extracellular markers, such as horseradish peroxidase and ferritin, show that the small tubular system which sometimes forms a network around some of the surface vesicles may be a form of reduced sarcoplasmic reticulum which is not part of the extracellular space. Elements of this so-called reticular network approach the surface membrane to within a

10 nm gap, and this gap may be traversed by electron-opaque material. Somlyo (1972), on this evidence, has suggested that this reduced reticulum may play an important role in EC-coupling. There is also some evidence that microsomes isolated from smooth muscle (possibly from the reticular network) do bind calcium.

However, we must bear in mind that a sarcoplasmic reticulum, as we understand the structure from skeletal muscle work, is definitely not present in smooth muscles, and it seems most unlikely that the minute fraction of reticulum present in some smooth muscles is capable of modulating calcium levels to the extent required to explain contraction and relaxation. In the case of arthropod visceral muscles, where a well-developed sarcoplasmic reticulum is present, it is most probable that the mechanism outlined above may operate. Clearly some sort of calcium storage site must be present within the visceral-muscle fibre, but the level of calcium storage may not be as high as seen in skeletal muscle. Syson (1974) has shown that, in a potassium-induced contracture of ileal smooth muscle, the calcium required for the initial phasic part of the contracture may come from immediate intracellular sources, but the calcium required for the tonic sustained part of the contracture may come from extracellular sources via the plasma membrane, since the calcium conductance of the latter is related to the level of membrane depolarization. This sort of evidence suggests that the level of intracellular calcium storage may not be very high.

Examination of electron micrographs of many visceral muscles reveals the presence of numerous surface micropinocytotic vesicles. These begin as small inpushings of the plasma membrane, and at a later stage they separate and enter the outer part of the sarcoplasm just below the plasma membrane. The contents of these vesicles, rather like the contents of the T-system, are similar to the extracellular medium—a medium rich in calcium compared with the myoplasm. A complication to be considered also is that mitochondria may be found in very close association with the surface vesicles, the gap distance between the vesicles and the mitochondrial outer membrane often being as small as 4–5 nm. This observation has raised the possibility that either the surface vesicles or the mitochondria, or both, may be involved in the process of calcium release and uptake associated with contraction and relaxation.

Hurwitz et al. (1973) have shown that guinea-pig ileal smooth muscle 'microsomes' actively sequester calcium in the presence of magnesium and ATP. From the method of preparation it is not clear just what cell structure these particles were derived from—they could well be mitochondrial, reticular, or surface vesicular in origin. In a more controlled study, Syson (1974) prepared vesicle preparations from separate mitochondrial and plasma membrane fractions, in order to follow calcium uptake and release. The fine structure of the plasma membrane vesicle fraction can be seen in figure 5.7. It is important in studies of calcium uptake to be sure that it is active ATP-promoted uptake which is being studied. At an external calcium concentration of 2 mM, Syson (1974) found that there was an approximate two-fold increase in calcium uptake by the plasma membrane vesicular fraction in the presence of magnesium and ATP compared with blank controls, but little difference in uptake in these two conditions using mitochondrial fractions. This strongly suggests that any calcium uptake by mitochondria may be due to simple passive diffusion of Ca^{45} into the mitochondrial fractions, followed by electrostatic binding. Some calcium uptake may take place in mitochondria, but the level is too small to account for the calcium requirements of the cell in EC-coupling.

Figure 5.10 shows calcium uptake by both mitochondrial and plasma membrane vesicular fractions from rat ileal smooth muscle in different calcium media. It is immediately clear that the plasma membrane fraction possesses considerable powers of calcium sequestration compared with the

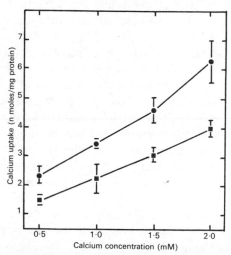

Figure 5.10. Calcium accumulation by plasma membrane vesicular fractions (closed circles) and mitochondrial fractions (closed squares) isolated from rat ileal smooth muscle. Note the considerable powers of calcium binding by the plasma membrane fraction, and the steep increase in binding as calcium concentration is raised in the presence of ATP. Courtesy of Dr. A. J. Syson.

mitochondrial fraction. It is difficult to say whether this fraction alone possesses enough binding capacity to explain cellular modulation of calcium during EC-coupling. A critical test would be to see whether this vesicular fraction was able to relax isolated smooth-muscle actomyosin, but this has not so far been investigated. Without a doubt, mitochondria play some role in calcium translocation during EC-coupling in visceral muscles, but this role may be only a general 'mopping up' during relaxation—the greater binding tenacity of the plasma-membrane vesicular fraction indicates that it is this fraction which may be responsible for the ultimate pumping of sarcoplasmic calcium down to levels low enough to permit relaxation. It must be remembered, however, that different visceral muscles differ in the relative balance of mitochondria/surface vesicles, and in some muscles the mitochondria may play a more significant role than that found in rat ileal muscle. Mitochondria in skeletal muscles certainly avidly bind calcium, and this may be the case in some visceral muscles, but this problem has not yet been thoroughly investigated. For an alternative view see Somlyo (1972).

So far, no studies have been carried out on calcium uptake by what could be regarded undisputably as smooth-muscle 'sarcoplasmic reticulum', and so it is not possible at this stage to be sure whether any reticulum present may also bind and release calcium. What emerges from the work mentioned above is that the terminal stages of calcium control in EC-coupling are far from being solved as yet. It may well be that calcium storage within visceral-muscle cells is low, and that the binding activity of 'reticulum' or plasma-membrane vesicles or mitochondria, or all of these agencies, may be adequate to reduce myoplasmic calcium to explain relaxation satisfactorily, but this is still speculation at the moment.

Modification of excitation-contraction coupling by drugs

Since EC-coupling is such a circuitous process involving a series of interdependent stages, a wide variety of physical and chemical treatments can disrupt or modify coupling by altering one or more of the stages in the coupling cycle. The most obvious points at which EC-coupling can be modified are:

    (1) Modification of the action potential, or the actual level of the mechanical threshold.
    (2) Modification of the inward transmission of excitation at the level of the T-system.
    (3) Disruption of the sarcoplasmic reticulum structure.
    (4) Modification of calcium release and uptake by the SR by agents which release or compete with calcium for binding sites.

Essentially, points 2 and 3 are purely physical phenomena. Mention has

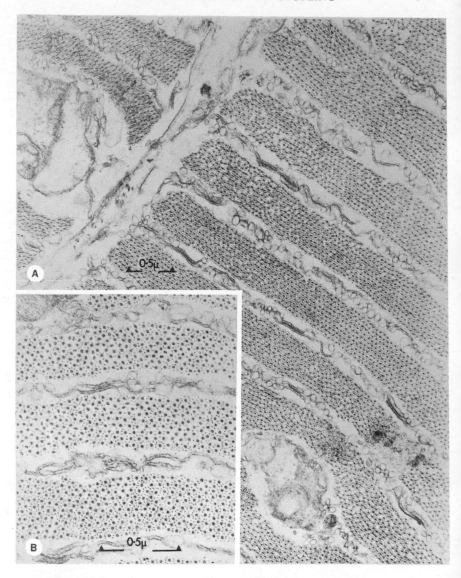

Figure 5.11. The effect of parathion (0·5 mM) on the fine structure of the sarcoplasmic reticulum of cockroach skeletal muscle. A, Whole field of myofibrils and SR. B, Higher power view of SR. Notice the almost complete vesiculation of the whole reticular network, while the T-system is relatively unaffected by this treatment. Author's original micrograph.

already been made briefly of glycerol-induced osmotic destruction of the T-system (Howell, 1969). This treatment leaves the later stages of EC-coupling unaffected, since Sakai *et al.* (1970) were still able to record perfectly normal caffeine contractures in muscles treated in this way. This type of treatment simply acts to divorce the plasma membrane depolarization from the calcium-releasing agencies within the cell. Recently it has been found that some chemical agents can selectively disrupt the fine structure of the sarcoplasmic reticulum. Caffeine has this sort of action on skeletal muscle (Huddart and Oates, 1970; Borys and Karler, 1971) as does parathion (Bradbury, 1973*b*) and DDT (Huddart *et al.*, 1974). The result of this action can be seen in figure 5.11. The physiological result of this is a massive rise in myoplasmic calcium followed by intense contractile activity. However, by far the most important agents which alter coupling are those which modify the mechanical threshold or modify calcium modulation processes within the muscle fibre. These agents fall into three well-defined categories:

(1) Anions such as bromides, nitrates, iodides, thiocyanates and methyl sulphate, which largely act to lower the mechanical threshold of muscle fibres, thus enhancing contraction. There is now some evidence that these agents may also affect calcium binding by sarcoplasmic reticulum.
(2) Heavy metal cations such as zinc, uranyl acetate, strontium, cadmium and barium. These act mainly to prolong the action potential, thus prolonging the active state and enhancing contraction. Again there is evidence that these cations may modify calcium binding by the sarcoplasmic reticulum, causing a release of calcium into the sarcoplasm.
(3) Alkaloid drugs such as caffeine, quinine, quinidine, nicotine and ryanodine. These have a variety of actions, both on the mechanical threshold and on calcium binding by the sarcoplasmic reticulum.

As with so many other phenomena in muscle physiology, studies have been largely confined to skeletal muscle, and it is only in recent years that any attention has been paid to visceral muscle. Since next to nothing is known about anion and divalent cation action on visceral-muscle EC-coupling, our attention here will be centred on the alkaloids about which something is now known in relation to visceral muscle. For an overall review of EC-coupling modification in skeletal muscle by the above list of agents, the excellent review of Professor Sandow (1965) should be consulted.

The main alkaloids which modify contractile output of skeletal, cardiac and visceral muscle are caffeine, quinine, nicotine and ryanodine, these being quite unrelated substances.

The first obvious major action of caffeine in causing contraction en-

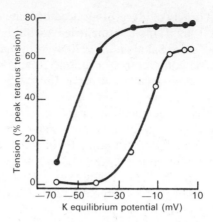

Figure 5.12. The effect of 2 mM caffeine on the tension/depolarization curve of locust skeletal muscle. Open circles represent control conditions, closed circles caffeine salines. Notice the significant lowering of the mechanical threshold in caffeine salines. Redrawn from data in Huddart and Abram (1969).

hancement in skeletal muscle is at the trigger stage of EC-coupling. Although caffeine does not significantly affect the value of the membrane potential, it does cause a great lowering of the mechanical threshold. If tension is plotted against external potassium or membrane potential, a tension/depolarization curve is obtained showing the presence of the mechanical threshold. If this experiment is repeated in the presence of caffeine, a second curve is obtained, and two such curves are shown for insect skeletal muscle in figure 5.12. It will immediately be apparent that caffeine has lowered the mechanical threshold from about -28 mV to about -50 mV. Depressions of mechanical threshold of this order have been recorded in mammalian, amphibian, crustacean and insect skeletal muscle.

When similar experiments are carried out on visceral smooth muscle, rather surprising results are obtained. Syson (1974) found that addition of caffeine to rat ileum caused an abolition of both spikes and slow waves, and an inhibition of spontaneous contractions (figure 5.13). Caffeine also considerably depressed KCl-induced contracture tension of rat ileal and vas deferens smooth muscle. When tension is plotted against depolarization (i.e. potassium concentration), a significant shift of the curve to the *right* is seen. Far from enhancing contraction by lowering the mechanical threshold, caffeine has the reverse effect on visceral muscle, causing a depression of responsiveness.

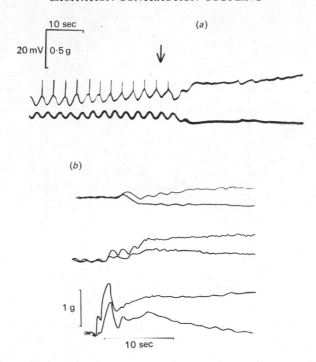

Figure 5.13. (*a*) The effect of 5 mM caffeine on the spontaneous electrical and mechanical activity of rat ileal smooth muscle (added at arrowed point).
(*b*) Depression of KCl-induced contractures in ileal muscle by the addition of 2 mM caffeine with the KCl saline. Upper two records, 15 mM KCl; middle records, 30 mM KCl; lower records, 50 mM KCl. In all cases the upper of the two traces is the control contracture, the lower trace the same preparation with KCl-caffeine saline. Courtesy of Dr. A. J. Syson.

The second major action of caffeine on skeletal muscle is at the level of the sarcoplasmic reticulum, where caffeine affects calcium-accumulating ability (Weber, 1968; Carvalho, 1968*b*). Carvalho (1968*b*) demonstrated that the sarcoplasmic reticulum possessed two fractions of calcium, a large non-labile fraction (about 80% of total calcium), and a smaller labile fraction (about 20% of total calcium), actively bound in the presence of ATP. It is this latter calcium fraction which is cyclically released and bound to activate and terminate contraction. Carvalho showed that this labile calcium fraction was strongly affected by caffeine, caffeine action being to release this calcium, causing massive activation of the contractile system. In a most elegant experimental study, Weber and Herz (1968)

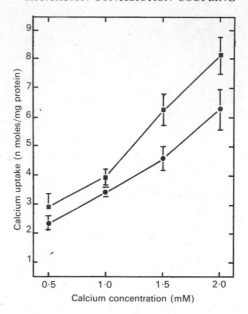

Figure 5.14. The effect of 5 mM caffeine on calcium uptake by plasma-membrane vesicular fractions isolated from rat ileal smooth muscle. Squares represent uptake in caffeine incubation media, circles represent uptake in control media. Each point is the mean ± S.E. ($n = 6$). Note the enhancement of calcium uptake in the presence of caffeine. Redrawn from data in Syson (1974).

successfully correlated this calcium-releasing action of caffeine on reticulum with its action on the intact muscle fibre.

Syson (1974) has investigated the effect of caffeine on calcium accumulating ability of membrane vesicular fractions isolated from rat ileal smooth muscle. The results of this study can be seen in figure 5.14. Far from releasing calcium from this membrane fraction, caffeine was found actively to *promote* calcium binding, resulting in a fall in myoplasmic free calcium and an inhibition of contraction. That this action of caffeine on the plasma-membrane fraction correlates with the inhibition of contraction is further shown by Ca^{45} efflux experiments. When isolated longitudinal ileal smooth-muscle strips are loaded with Ca^{45}, and efflux has proceeded to the slow sustained component related to intracellular calcium, an addition of caffeine was found to cause an *inhibition* of efflux (Syson, 1974, figure 5.15). This is the reverse of the situation in skeletal muscle where caffeine stimulates calcium efflux (Isaacson, 1969). Not only was caffeine found to promote calcium binding by the plasma-membrane

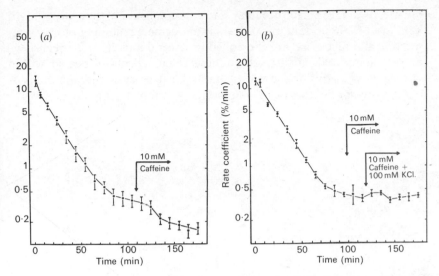

Figure 5.15. (*a*) The effect of 10 mM caffeine saline on the Ca^{45} efflux rate coefficient of rat ileal smooth muscle. Note the reduction in efflux caused by caffeine.
(*b*) The effect of 20 min pretreatment with 10 mM caffeine on the potassium-induced increase in Ca^{45} efflux rate coefficient of rat ileal smooth muscle. Note the inhibition by caffeine of the normally massive stimulation of calcium efflux by KCl. Redrawn from data in Syson (1974).

fraction and inhibit calcium efflux from whole muscle, prior treatment with caffeine also inhibited the expected stimulation of Ca^{45} efflux by subsequent KCl application (figure 5.15).

Quinine is the second major alkaloid known to have rigour-promoting effects on skeletal muscle (Isaacson *et al.*, 1970; Huddart, 1971), and these effects have been shown, like those of caffeine, to be related to an inhibitory action on calcium sequestration by the sarcoplasmic reticulum. On visceral muscles, quinine again has a *relaxatory* action, and Syson (1974) has been able to relate this action to quinine-promoted increase in plasma-membrane vesicular fraction binding.

Alkaloid drugs have proved a most useful tool in analysing the importance of various stages of the EC-coupling cycle in skeletal muscle, particularly in highlighting the extreme importance of the intracellular calcium modulation process in switching contraction on and then off. These drugs have proved to be no less important in revealing the EC-coupling process in visceral muscle, even though their effects are the exact reverse of what was expected in the light of previous work on skeletal muscle. By modifying intracellular calcium mobilization, these drugs highlight just how important this process is in our understanding of how contraction and relaxation are controlled in visceral muscle. It is because we as yet understand so little about how calcium is translocated within visceral muscles during the contraction-relaxation cycle that our analysis of EC-couplings in these strange muscles is so imperfect and tentative.

CHAPTER 6

THE MECHANICAL ACTIVITY OF VISCERAL MUSCLE

Studies on the mechanical activity of muscle have been largely confined to skeletal, and to a lesser extent cardiac, muscle. Although visceral muscles have been largely ignored, many of the major concepts of muscle mechanics also apply to visceral muscles. In this chapter, some of the more important aspects of mechanical activity will be considered, in as much as they can be applied to visceral muscle.

Muscles can be regarded as biological machines which convert chemical energy into mechanical energy, but various muscles vary greatly in the efficiency of this conversion. The efficiency of conversion is affected by factors operating at the intracellular level, i.e. in the internal conversion of chemical energy into myofibrillar force, and also at the whole muscle level in the mode of expression of this force against an external load. Of the three kinds of muscle: skeletal, cardiac and visceral, the latter is regarded for a variety of reasons as being the least efficient in the conversion of chemical energy into mechanical energy.

In skeletal muscle, the orientation of the individual fibres within the muscle, and their orientation with regard to the insertions on the skeleton are of crucial importance in determining the efficiency with which the contractile outputs of the individual fibres are transmitted against the load. Muscles with long straight fibres running directly in the plane of the insertions, such as the biceps or sartorius, are by far the most efficient in transmitting tension against a load. On the other hand, muscles with small diffuse radiating fibres, such as the deltoid, will have a lowered directional exertion of tension.

Visceral muscles which, with the exception of arthropod gut muscles, are largely organized as multi-directional overlaid sheets composed of irregular fibres (see figures 6.1 and 6.2) must be regarded as far less efficiently constructed in terms of their directional organization of tension

111

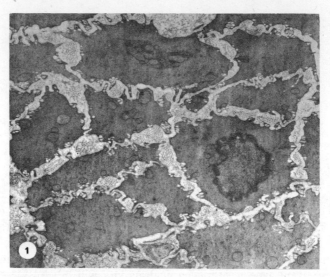

Figure 6.1. Low-power electron micrograph of rat ileal smooth muscle, showing the great irregularity of the fibres and the lack of directional fibre organization. Print magnification × 7,500. Author's original plate.

exertion. An additional important factor to be remembered is that visceral muscle cells often form only a small part of many of the organs in which they operate (e.g. in the gut, the alveoli, lymphatic vessels, the prostate and urethra). This means that these visceral muscles are surrounded by a large quantity of non-contractile tissue, which forms a considerable hindrance to the efficient development and maximization of contractile force. This will be immediately apparent from the low-power photomicrograph of mammalian gut in figure 6.2, where outer sheets of visceral muscle cover vast quantities of secretory tissue.

When considering mechanical efficiency, it must be remembered, however, that visceral-muscle cells comprise the walls and operative parts of complex hollow organs, where a multi-directional exertion of tension is required, and these muscles must thus be regarded as mechanically efficient and ideally constructed for their particular function. In addition, many visceral muscles can maintain tension with only a small increase in oxygen consumption, indicative of high efficiency of energy conversion into tension. These muscles are only inefficient in absolute terms when compared with skeletal muscle where there is a maximization of energy conversion into direct tension. In the case of skeletal muscle, it is possible to compare their gross mechanical efficiency in terms of their

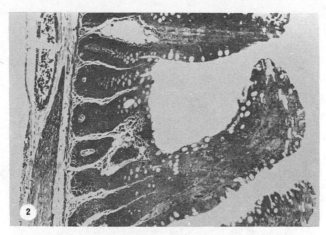

Figure 6.2. Light micrograph of rat ileum showing the circular smooth-muscle layer. The outer longitudinal layer has been removed. Note the small proportion of the whole tissue which is muscular. Print magnification × 150. Plate courtesy of Dr. A. J. Syson.

leverage. Since skeletal muscles exert tension on a load over a joint, they act as levers, the relationship between the position of the force, the load and the joint determining whether muscles are first, second or third-order levers. With visceral muscles, there is no consistent relationship between the point at which tension is exerted and the load itself, so lever factors cannot be readily applied here.

So far we have considered only gross mechanical efficiency. The internal efficiency of a muscle fibre, expressed in terms of maximizing tension output per unit cross-sectional area of the fibre, is dependent upon two main factors: (1) The volume of the contractile component in relation to the volume of the non-contractile component of the fibres, and (2) the disposition of both the myofibrils within the fibre and the myofilaments within the myofibrils themselves. Here we come across a major problem in dealing with visceral muscles. In skeletal and cardiac muscles, the contractile machinery of the fibres is organized into distinct myofibrils, but the contractile machinery of visceral muscles (with the exception of the arthropods) is not organized into distinct myofibrils or any other obvious cell subdivision. This can be clearly seen when rat ileum, for example (figure 6.1) is compared with cockroach flight muscle (figure 6.3). Whereas in skeletal and cardiac muscle the basic unit of the contractile component is the myofibril, divided into serial sarcomeres, in visceral muscle it is the myofilament, and myofilaments can be seen

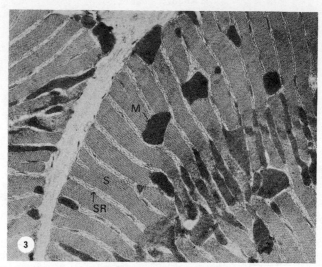

Figure 6.3. Low-power electron micrograph of cockroach flight muscle, showing the orderly arrangement of the contractile component into columnar sarcomeres (S) separated by tubules of the sarcoplasmic reticulum (SR) and mitochondria (M). Print magnification × 7,500. Plate courtesy of Mr. M. Greenwood.

scattered throughout the sarcoplasm, with little semblance of order. Not only is myofilament packing poorly organized, there is also little semblance of any ordered relationship between the thick and thin filaments (even in those muscles where thick filaments can be seen at all). This can be seen in figure 6.4, where the myofilament disposition in a typical visceral muscle is compared with that in a skeletal muscle. In all muscles, tension is generated as a result of cross-bridge formation between the thick and thin filaments. Since the myosin filament possesses six cross-bridge sites per revolution of 'pitch' of the filament, it is clear that a neat orbit of six actin filaments around one myosin filament will be optimal for maximum generation of myofilamentary force. Departure from neat 6:1 actin:myosin orbital arrangements will of necessity lead to some inherent loss of contractile force, and this has been most elegantly demonstrated for a variety of skeletal muscles by Auber (1967). As can be seen from figure 6.4, there is no evidence of neat hexagonal actin:myosin orbital arrangements in visceral muscle, and it is clear that myofilamentary shortening force and capacity will be relatively inefficient in these muscles compared with that in most cardiac and skeletal muscles. For further details concerning the contractile apparatus and its organization in visceral muscle, see Chapters 1 and 7.

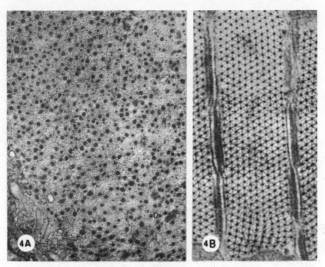

Figure 6.4. (A) Myofilament organization in the smooth muscle of *Buccinum* pharyngeal retractor. Notice the lack of myofibrils and the lack of clear actin/myosin orbits. Print magnification × 25,000. Author's original plate. (B) Cockroach flight muscle myofilament system, with clear myofibrils and hexagonally-arrayed myosin filaments with neat 6-membered actin orbits. Print magnification × 37,500. Author's original plate.

Since a muscle fibre possesses both a contractile and a non-contractile component, the actual volume of the contractile component is of great importance in terms of maximizing the tension output of the fibre. The force developed by the interacting myofilaments must be transmitted against the load through the sarcoplasm and the relatively gelatinous extracellular matrix. The reason that the non-contractile component is of such importance is that neither it nor the extracellular matrix is inextensible. Part of the myofilamentary force has to be spent in overcoming the extensibility of the non-contractile component, such as the elasticity of the sarcolemma, and the viscosity of the sarcoplasm and its inclusions, which represent a considerable fluid damping.

This situation can best be appreciated in a simple mechanical diagram of the muscle cell (figure 6.5). It can be seen that the myofibrillar force acts against the load through a set of series and parallel elastic components, and its action is damped by sarcoplasmic viscosity. The very presence of these other components of the cell ensures that only a partial conversion of the energy of the contractile component takes place in the form of force acting against the external load. From the mechanical diagram it will be clear that the larger the contractile component, the less will be the drag and damping, and the greater will be the overall fibre efficiency. It will thus be obvious that visceral muscles, with their large non-contractile components, are inherently inefficient. However, the possession of a large non-contractile component is important in conferring the property of extensibility on muscle. Visceral-muscle fibres, because of this component, act as strong elastic bodies, and they possess enormous resistance to stress. This is a most necessary property, since many visceral muscles (such as in the gut and the vascular system) are called upon in normal operative conditions to resist much greater extension than that possible for skeletal muscle.

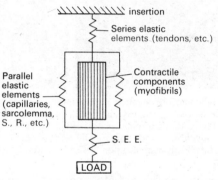

Figure 6.5. Simplified conceptual diagram to illustrate some of the mechanical properties of a muscle fibre.

Central to our whole understanding of contractile activity in muscle is the concept of *active state*, and this is an exceedingly difficult pheno-menon to define in any easily meaningful way. The earlier definitions of Hill (1950) and Mommaerts *et al.* (1961) define active state solely in terms of the activity of the contractile component. These definitions are of little help when we consider the muscle cell as a whole system. A more physiologically meaningful definition of active state in terms of the con-traction of the muscle fibre is to say that as long as the contractile elements are shortened from the resting condition (i.e. as long as the actin and myosin myofilaments are interdigitated more than at rest), then the muscle fibre is in the active state. This definition (largely related to work on vertebrate skeletal muscle) appears simpler than it really is. Contraction and relaxation are merely measures of the transient intensification and reduction of the active state (i.e. measures of greater or lesser degrees of actin (myosin interaction). We now know that even at rest, some cross-bridges still connect the myosin and actin filaments. It is doubtful if any muscle cell has a true 'resting condition', since the number of patent cross-bridges at any one relaxatory event may be quite variable. While the active state may be difficult to define, it is of profound importance in relation to the non-contractile component.

At first it is difficult to appreciate that a muscle fibre is in the active state, since the active state develops within a few milliseconds following stimulation, and its time course may only last about 25 msec. These events occur without visible shortening of the fibre, i.e. under truly isometric conditions, the only indication that something is actually happening being the liberation of heat. Development of the active state is accompanied by an abrupt increase in heat production—the *activation heat*—and since this happens in isometric conditions, it is independent of the load on the muscle. Activation heat is a measure of the time course of the active state, and it relates to the energy production due to shortening of the myofibrils, or the interdigitation of the myofilaments in visceral muscles. Subsequent to activation heat the fibre shortens, and this is accompanied by a further phase of heat production, the slow-rising *shortening heat*. The latter relates to the compression of the non-contractile component and the action against the load, and hence it is load-dependent. It will be clear from this description that the duration of the active state (indicated by activation heat) and the time course of the muscle fibre contraction do not coincide. A fuller account of these topics can be found in the general textbook by Florey (1966).

Since the contractile component must overcome the elasticity and damping of the rest of the cell before the action on the load takes place, there is a considerable delay before shortening occurs. The active state not only occurs under isometric conditions, but it is well on the decline before visible fibre shortening occurs. In many cases the muscle fibre is maximally shortened when the contractile elements are back in the resting condition and the active state has decayed completely. The time course of these events is illustrated diagrammatically in figure 6.6.

The important consequence of these events is that following a single stimulus, the active state is of insufficient duration to overcome fully the drag of the non-contractile component. This means that in a single twitch the muscle is unable to shorten to its maximum possible extent, and it does not develop maximal tension under twitch conditions. With repetitive stimulation, the active state is constantly re-primed to maximal intensity, allowing eventual overcoming of the non-contractile component. Hence, when stimulus rate is sufficient to maintain the active state at a constant maximal level, the fibres shorten maximally and develop maximal

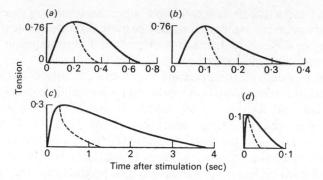

Figure 6.6. Curves displaying the time course of the twitch (solid lines) and active state (broken lines) in (a) frog sartorius muscle; (b) locust flight muscle; (c) *Mytilus* pedal retractor muscle; and (d) cicada tymbal muscle. Redrawn from Florey (1966).

tension against the load. It is for this reason that tetanic tension is always higher than twitch tension. These concepts can be seen in diagrammatic form in figure 6.7. It will be seen that the tetanus/twitch ratio of a particular muscle is a measure of two factors: (a) the speed with which maximal active-state conditions can be established, and (b) the relative volumes of the contractile and non-contractile components. In fast skeletal muscles, the active state always achieves maximal intensity with each stimulus. Here it is only the *duration* of the active state which is altered by repetitive stimulation. In these muscles, the tetanus/twitch ratio is small, varying between 1·5 and 2·5, for example, in frog muscle. In fast skeletal muscles with a high tetanus/twitch ratio, as in some insect muscles, for example, the ratio simply reflects the slowness of establishing maximal maintained active-state conditions. In slow skeletal muscles we are faced with a different problem, since here the active state does not develop to maximal intensity as the result of a single stimulus. Here, repetitive stimulation causes both a progressive summation to maximal active-state conditions, and then a protraction of these maximal conditions. This is shown conceptually in figure 6.7. As a result, slow muscles usually require a high stimulus frequency to achieve maximal tension, the relationship between stimulus frequency and tension is almost linear and tetanus/twitch ratios are usually very high.

In visceral muscles, there is a large non-contractile component, and an enormous gelatinous extracellular space filled with collagen with a high degree of passive elasticity. There is a further complication, however, in relation to the so-called visceral-muscle 'twitch'. As we understand the term 'twitch' in skeletal muscle, there are few visceral-muscle equivalents.

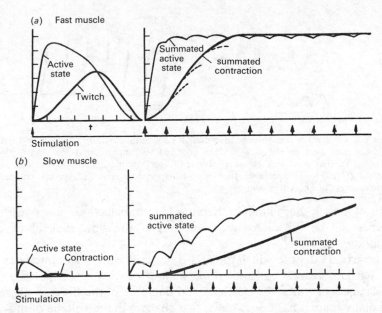

Figure 6.7. Conceptual graphs of the duration of the active state and mechanical activity during the twitch and tetanus in fast and slow skeletal muscle. Note how in slow muscle the active state must itself summate before significant tension is generated. Adapted and redrawn from Florey (1966).

Most visceral muscles respond with contractions only to trains of electrical stimuli, the stimuli being spontaneous spikes in rhythmic muscles and bursts of neurally induced action potentials in non-rhythmic muscles. Figure 6.8 shows typical mechanical activity in two different visceral muscles. In the case of the ileum, a burst of action potentials is needed to elicit a single slow wave of contraction, while in the vas deferens a burst of applied stimuli at high frequency is required to elicit a response. It is this latter type of response which has often been called a 'twitch', but this term is clearly inappropriate. It must be remembered, however, that some visceral muscles do respond to a single stimulus with a twitch, but such a situation is rather rare. The need for repetitive stimulation to induce contractions or any really noticeable tension development in visceral muscle is a measure of the very slow development of the active state in these muscles, the active state being probably of even lower intensity following a single stimulus than that of slow skeletal muscle. It would seem that an intense sustained shortening of the myofilamentary contractile system is needed before any manifestation of this is apparent in

(a)   (b)

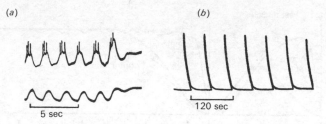

Figure 6.8. Contractile responses of mammalian visceral muscle. (a) Rat ileum. A burst of spikes superimposed on the slow wave is required to generate rhythmic contractions. (b) Rat vas deferens 'spike-like' contractions due to direct stimulation every 60 sec for a 1-sec burst at 100 Hz. This considerable stimulation is required to generate appreciable tension. Figure courtesy of Dr. A. J. Syson.

terms of muscle shortening. In their mechanical behaviour, most visceral muscles can be considered as extensions of the slow skeletal-muscle condition.

The actual extent to which the non-contractile component interferes in the tension generation and transmission process can be seen in experiments in which the elasticity and damping of the fibres is altered. For many years it has been known that changes in the volume of muscle fibres, brought about by changes in the osmotic strength of the bathing medium, greatly modify tension output (Fenn, 1936; Hodgkin and Horowicz, 1957). When muscle cells are exposed to hypo- and hyperosmotic bathing media, they behave in varying degrees like osmometers, skeletal muscles being almost perfect osmometers, while cardiac and visceral muscles are progressively poorer osmometers (Bozler, 1965). The swelling of muscle cells in hypotonic media generates a considerable internal hydrostatic pressure which acts to overcome much of the elasticity and damping of the cell. This has profound consequences in terms of contractile output, since the active-state event (i.e. the shortening of the myofilaments) is far more effective in inducing fibre shortening which results in enhancement of tension (Baskin, 1967). In visceral muscles, where the non-contractile component is large, hypotonic media cause even greater enhancement of tension than that seen in skeletal muscle. This can be seen in figure 6.9, where responses of rat ileum to KCl-induced depolarization are shown over a wide range of bathing-medium tonicities. Maximal KCl contractures were recorded in the range 70–80% normal saline tonicity, these contractures being as much as 30% higher than the mean control value. Grossly hypertonic media, which shrink the cells and distort the extracellular matrix, depress KCl contractures. Similar effects of tonicity changes on cell volume and spontaneous

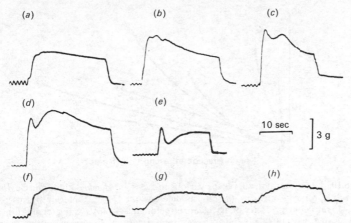

Figure 6.9. Effect of tonicity of the saline on 50 mM KCl contractures of rat ileal smooth muscle. Tonicities were as follows: (a) 100%, (b) 90%, (c) 80%, (d) 70%, (e) 60%, (f) 100%, (g) 120%, (h) 140%. Hypertonic salines reduce and hypotonic salines enhance contracture tension. Calibrations 10 seconds and 3g. From Syson and Huddart (1973).

activity have been reported in vascular smooth muscle (Jonsson, 1969), the rat uterus and vas deferens (Carroll, 1969).

A further, and somewhat surprising, effect of hypertonic salines on muscle cells is a distinct increase in heat production and oxygen consumption. Both phenomena cannot be accounted for by any increase in resting metabolism; they appear to be related to some form of activation of the contractile system. In skeletal muscle, the sarcoplasmic reticulum becomes distended in hypertonic salines (Birks and Davey, 1969) and this almost certainly results in some release of reticular calcium into the myoplasm, causing an increase in myosin ATPase activity. This increase in ATPase activity is the probable cause of heat production and increased oxygen consumption, but it does not become manifest in the form of fibre shortening, since the increase is small and the hypertonic fibres are shrunken and exhibit reduced responsiveness. In the case of visceral muscles, where there is little or no sarcoplasmic reticulum (see Chapter 1), hypertonic salines, which cause increased internal ionic strength, may induce release of some of the vesicle-modulated calcium store (see Chapter 5), still resulting in increased myofilamentary activity.

Length/tension relationship

Owing to the presence of the non-contractile component, a resting muscle has many of the properties of an elastic body, and is able to resist

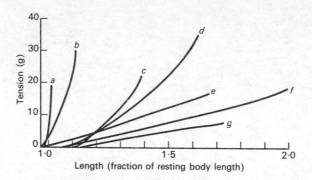

Figure 6.10. Length/tension curves for various muscles. (*a*) *Bombus* flight muscle, (*b*) locust flight muscle, (*c*) frog sartorius, (*d*) rat vas deferens, (*e*) *Mytilus* anterior byssus retractor muscle, (*f*) rat ileum, (*g*) *Helix* pharyngeal retractor muscle. (*a*), (*b*), (*c*), (*e*) and (*g*) redrawn from Hanson-Lowy (1960), (*d*) and (*f*) from original data. Note how extensible the visceral muscles are (*d*, *e*, *f* and g) compared with skeletal muscle.

compression and stretch by developing a mechanical resistance, known as *passive tension*. This elastic tension increases in a non-linear manner as the muscle is progressively stretched, there being little passive tension at first but, as the muscle is further stretched, the tension rises exponentially with linear increase in applied stretch. The relationship between muscle length and tension can be seen plotted for a variety of muscles in figure 6.10. The steepness with which muscles develop tension on stretching can be seen to vary widely, being exceedingly steep in insect flight muscle, less steep in skeletal muscles, and very sloping in visceral muscles. Difference in abruptness of passive-tension development correlates well with differences in the volume of the non-contractile component. Vas deferens muscle, for example, develops passive tension more abruptly than intestinal muscle, the latter having a higher extracellular space and lower contractile component volume. The length/tension relationship is simply a correlate of functional differences between muscles. Flight muscle, of insects for example, is organized to develop massive tension with little alteration in sarcomere length (little more than 2% shortening). This type of muscle has great resistance properties to stretch and normally alters little in length. Normal skeletal muscles of vertebrates usually shorten (when maximally activated) to about 65–70% of resting length. Visceral muscles, on the other hand, have enormous resistance to stretch, and are not damaged during enormous length changes.

As a muscle is stretched, the elastic components are stretched, enabling the contractile system to exert its shortening force against the load more effectively when the muscle is stimulated, a condition similar to that in muscles treated with hypotonic solutions. Hence, as a muscle is stretched, the larger is the recorded isometric tension following stimulation. Total isometric tension is thus a function of muscle length, but this tension is composed of two components, the original *passive* tension and the subsequent *active* tension. The latter is simply the tension which develops on activation of the muscle. Total isometric tension is, within wide limits, a continuously rising function of stretch (i.e. muscle length), but active tension is not. The reason for this is that as the muscle is progressively stretched, passive tension rises so greatly that it may eventually exceed active tension, and as a result, active tension may decline with stretch.

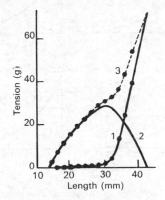

Figure 6.11. The relationship between passive tension (1), active tension (2) and total tension (3) and length of rabbit uterine smooth muscle. Redrawn from Csapo (1960).

This situation can best be appreciated by examining the length/active tension curve for uterine smooth muscle in figure 6.11. Here, muscle length has been plotted against changes in passive, active and total tension. Although total tension does rise, active tension eventually falls. In this situation, stimulation of the muscle may result in a transient fall in tension as the cross-bridges begin to open at the beginning of the bridge-formation cycle.

Another phenomenon in muscle mechanics related to length/tension characteristics is the force/velocity relationship. Under isotonic conditions, as the load on a muscle is increased, velocity of shortening decreases. When external load on a muscle is plotted against corresponding velocity of shortening, a force/velocity curve is obtained. Figure 6.12 shows force/velocity plots for three smooth muscles compared

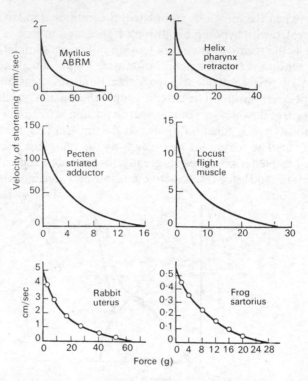

Figure 6.12. Force/velocity curves for some skeletal and visceral muscles. Redrawn after Florey (1966).

with a plot for a skeletal muscle, and it can be seen that smooth muscle differs little from skeletal muscle in this mechanical property. It can be seen that decline in shortening velocity is not a linear relationship with increase in load, and this is something of a mystery since the rate of energy liberation during contraction increases as the load increases. In other words, it is the load on the muscle itself which determines the rate of energy-releasing reactions within the fibres (Hill, 1938), a postulation borne out by more recent chemical investigations on the energetics of muscle during contraction (Kushmerick et al., 1969; Kushmerick and Davies, 1969).

The force/velocity curves show two things: firstly, that even with a *minimal* load, a muscle shortens with a distinct limited velocity; and secondly, maximum force develops only under isometric conditions at zero velocity of shortening. Hill (1938) has described the force/velocity relationship by a simple equation in which:

$$V = \frac{(P_0 - P)b}{P + a}$$

where V is the velocity of shortening, P_0 the maximum tension the muscle can develop at zero speed (maximum isometric tension), P the load on the muscle, a is a constant in units of force, and b is a constant in units of velocity. This equation simply states that the velocity of shortening depends upon how much the maximal isometric force exceeds the load placed upon the muscle. Although a large number of different vertebrate and invertebrate skeletal and visceral muscles roughly follow this force/velocity relationship, there is a wide variation in the actual speed of shortening seen (see Table 6.1), and it can be seen that visceral muscles are right at the bottom of the list.

Table 6.1.—The maximum speed of shortening (V_0) of some selected skeletal and visceral muscles.

Animal and muscle	Temperature (°C)	V_0 (muscle lengths/sec)	Reference
rabbit uterus	37	0·2	Csapo (1955)
frog sartorius	22	10·0	Hill (1938)
tortoise illiofibularis	20	0·4	Abbott and Lowy (1957)
locust flight muscle	30	13·0	Hanson and Lowy (1960)
Pecten adductor (striated)	14	3·0	Hanson and Lowy (1960)
Pinna adductor (tonic)	18	0·1	Hanson and Lowy (1960)
Octopus funnel retractor	18	2·4	Hanson and Lowy (1960)
Helix pharyngeal retractor	14	0·2	Abbott and Lowy (1953)

Isometric and isotonic contraction

In this chapter, reference has been made to the terms *isometric* and *isotonic contractions*. Although these are terms which are traditionally thought of as referring to skeletal muscle, they do have their application to aspects of tension development in some visceral muscles, particularly the discrete muscles of the arthropod gut, many molluscs and the muscles of the vertebrate urino-genital system. For this reason, a brief discussion of these terms is of value at this point.

Following activation, muscles expend energy during contraction, but the energy may be expended in a variety of ways, such as in overcoming the resistance of the non-contractile component, in heat production, and in mechanical energy exerted against the load. The relative proportions of total muscle energy which are diverted into these various channels varies, depending upon the nature of the contraction. If a muscle is allowed to contract against a load and the muscle physically shortens during this process, this type of contraction is defined as an *isotonic* contraction, since actual tension rise in the muscle is small. Since fibre shortening will only be maximal when the load to be carried is minimal, only an arrangement which allows maximal fibre shortening can be truly regarded as isotonic. With this sort of idealized situation, although some frictional and heat energy are liberated, the mechanical energy exerted against the load actually moves the load, and this is by far the greatest fraction of total muscle energy expended.

On the other hand, if a muscle is allowed to contract against a load, but is prevented from shortening, no mechanical energy is expended in actually moving the load. This mechanical energy now takes the form of a massive tension rise in the muscle. As the contractile system deforms the non-contractile component, its energy is exerted against the non-moving load, and a large heat-energy dissipation takes place. A contraction of this sort, with no change in length of the muscle is called an *isometric* contraction. The basic concepts behind isometric and isotonic contractions can be more readily appreciated in the simple conceptual diagram in figure 6.13. It will be clear that a true isometric contraction

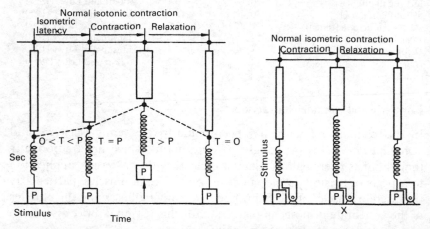

Figure 6.13. Conceptual mechanical diagrams to illustrate isotonic and isometric contraction. In isometric contraction, soon after the stimulus, load (P) is greater than tension (T) since the contractile component (cc) has not begun to shorten. As cc shortens and T exceeds P, the load is lifted. During this process the cc has deformed the series elastic component (Sec). With isometric contractions, the cc only acts to deform the Sec, exerting tension against the load. Adapted and redrawn from Sandow (1961).

is a hypothetical concept since such a contraction cannot be measured. Any recording device such as a wire strain gauge or piezoelectric transducer has to be at least minimally deformed by the muscle before it will register any tension, and this must involve some movement of the muscle fibres against the compliance of the recording device. Similarly, a truly isotonic contraction is also rather hypothetical since, even with a minimal load, some tension rise in the muscle is inevitable. Clearly isometric and isotonic concepts simply represent different ways of expressing and measuring the energy or force output of a muscle, otherwise they are of no fundamental importance in themselves. In the case of visceral muscles (as is true also of most skeletal muscles), contractile activity involves both changes in length and tension during the normal course of activity within the body.

Latency relaxation

In both skeletal and visceral muscles, it is a consistent observation that a small relaxation occurs after stimulation, during the latent period before the onset of positive tension generation (see Sandow, 1966). In the case of both types of muscle, but particularly in the case of visceral muscle, the greater the degree of stretch originally applied to the muscle, the greater is the latency relaxation. Largely due to the elegant work of Sandow (1966), most is known about latency relaxation in frog skeletal muscle, and this case will be used as the model for the phenomenon in this brief description. Figure 6.14 shows the typical onset and time course

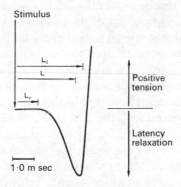

Figure 6.14. Tension changes in frog sartorius after stimulation showing latency relaxation and onset of positive tension. L_r, time to the onset of latency relaxation; L, time to the end of the latency relaxation; L_1, time to onset of positive tension. Redrawn from Sandow (1966).

characteristics of the latency relaxation of frog sartorius muscle. Latency relaxation in this particular preparation begins about 1 msec after the stimulus, reaches a peak at about 2·5 msec after the stimulus and is replaced by positive tension about 3·5 msec after the stimulus. In most cases, the latency relaxation is of the order of 2–5% total peak tetanic tension.

The cause of latency relaxation has been the subject of some debate. Sandow (1966) suggested two possible mechanisms: (1) a transient fall in myofilamentary resting tension; and (2) a decrease in tension caused by calcium release from the sarcoplasmic reticulum during excitation-contraction coupling. Both possible mechanisms are related to the sharp rise in myoplasmic free calcium during the later stages of the EC-coupling cycle. The work of Hill (1968) largely supports the former view, since he showed the existence of filamentary resting tension related to the presence of cross-bridges between actin and myosin filaments. This tension is small and, perhaps significantly, of the order of the observed latency relaxation in frog muscle. The work of Davies (1963) has already shown that calcium ions and the splitting of ATP are associated with cross-bridge breakage and reformation for the production of positive tension in muscle. It seems perfectly reasonable to propose that the increase in myoplasmic calcium following stimulation may cause an initial breakage of 'resting' cross-bridges, resulting in a transient fall in filamentary tension, before active reformation of cross-bridges takes place and active tension occurs.

An alternative view has been proposed by Sandow (1966) involving the calcium-releasing activity of the sarcoplasmic reticulum. It is now universally accepted that in skeletal muscle, the stimulus passes down the T-tubules to the triadic areas of the sarcoplasmic reticulum, causing a release of bound calcium by the lateral cisternae (see Chapter 5). These actions, and the diffusion of calcium to the myofilaments occurs during the latent period. Sandow has argued that it may be the actual calcium-releasing process which causes latency relaxation by causing osmotic changes in the sarcoplasmic reticulum. In this view, calcium release causes an osmotic-pressure decrease in the reticulum, causing water to leave the sarcoplasm and inducing a temporary tension decrease. This hypothesis is somewhat difficult to test, even though the time course of the diffusional events involved will adequately fit into the latency relaxation time course. A critical test would be to examine latency relaxation in muscles with very different reticulum volumes, a large relaxation being expected in muscles with a large reticulum volume.

In the case of visceral muscles, it is difficult to see how Sandow's view could be easily applied. Most visceral muscles possess little or no sarcoplasmic reticulum, and it is thought that inward calcium movement may be mediated by the sub-surface micropinocytotic vesicles. It is possible that the vesicles may swell due to calcium release, extracting water from the sarcoplasm, but this has not been examined in any detail. On balance, it seems more likely that visceral-muscle latency relaxation may be due to a transient fall in myofilamentary resting tension, but this view must also await confirmation by critical tests involving relaxation measurements at different stretch conditions.

CHAPTER 7

THE PROTEINS OF VISCERAL MUSCLE

Like all cells, those of muscle contain a vast diversity of proteins, some concerned with the organization and transport properties of their limiting and internal membranes, others concerned with mediation of their metabolic activities. The principal and most conspicuous proteins of the muscle cell are, however, those involved in the actual mechanics of the contraction process. These are proteins of a bifunctional character, part structural and part biologically active, organized into or assisting the organization of the long fibres or filaments which, as we have seen in Chapter 1, run through the major part of the cytoplasm of the cell along and parallel to its long axis. In skeletal muscle cells it is usual for these structural protein fibres to be organized into discrete functional units called *myofibrils*, the regular arrangement of which within the cell give it its characteristic banded appearance. In visceral muscle, however, there is not this type of easily recognized order, except in some specialized invertebrate situations, although the contractile protein fibres or filaments can usually be demonstrated by the appropriate electron-microscopical fixation and thin-sectioning technique.

It is perhaps not so surprising that the structural-contractile protein elements of vertebrate visceral-muscle cells are basically similar in composition and molecular structure to those of skeletal-muscle cells. As we might expect, however, there are differences in detail at the molecular level, reflecting the rather fundamental differences between the activities and physiological properties of visceral muscles, skeletal and cardiac muscles. Such differences are in addition to the more obvious differences in dispositions of the myofilaments *in situ*.

As is the case with skeletal and cardiac muscles, actin and myosin appear to be quantitatively the most important proteins of the contractile apparatus, with tropomyosin of slightly less quantitative significance. Most of the other proteins associated with the myofibril of skeletal

130

muscle appear also to be present in visceral muscle, although identification and characterization of these has not yet achieved the degree of sophistication applied to their skeletal-muscle counterparts. In addition the smooth muscles of many invertebrates contain important quantities of a protein called *paramyosin* which is not found in vertebrate visceral muscle.

The extractability of the contractile proteins

If the details of the molecular biology of the contraction process of a muscle type are to be elucidated, and hence some of the aspects of the physiology of contraction clarified, it will probably be necessary at quite an early stage to extract the proteins of the contractile apparatus of the muscle cell and purify them. This is in order that the various aspects of their composition, molecular organization, enzymology and interactive properties can be examined *in vitro* and correlated with the *in vivo* situation. The accessibility of the proteins of the contractile apparatus to selective extraction is therefore an important factor in the study of the biology of a muscle.

Vertebrate visceral muscle is frequently in a tissue situation where there is an extensive extracellular space containing relatively large amounts of insoluble components arranged in such a manner as to confer mechanical strength upon the tissue. Such insoluble material is most usually a network of collagen and elastin fibres embedded in a matrix of glycoproteins and acid mucopolysaccharides, all thought to be secreted by the muscle cells themselves. These insoluble components impede the disruption of the muscle cells by mild methods, and the subsequent release and extraction of the contractile protein. It is often necessary therefore to apply more drastic methods (such as grinding the tissue with sand to release the proteins) than would be applied to skeletal muscle, and this may bring with it doubts about the state of the proteins after extraction. This aspect of visceral-muscle biochemistry has been reviewed by Hamoir (1973).

The contractile proteins of visceral muscle, actin, myosin and tropomyosin, are extractable at low rather than high salt concentration, a situation in marked contrast to that found when contractile protein is being prepared from skeletal or cardiac muscle. Suitable low-ionic-strength conditions for extraction are, for example, at pH 7·4 in I 0·075 phosphate buffer. This difference in solubility properties between skeletal and visceral-muscle proteins is reflected in a very different partition of salt-extractable nitrogen for the two tissues when they are first extracted

Table 7.1.—Partition of salt-extractable nitrogen in various smooth and skeletal muscles extracted at neutral pH (from Hamoir, 1973).

	Smooth			Skeletal
	Human myometrium (non-pregnant adult)	Bovine carotid	Bovine inferior vena cava	Rabbit
% extracted at low ionic strength	80	61 86	69	33·4
% extracted at high ionic strength	20	39 14	31	66·6

with a low ionic strength and subsequently high ionic strength medium at neutral pH (Table 7.1). Often actomyosin not extracted at low ionic strength from smooth muscle can still be completely brought into solution at the same ionic strength by addition of ATP to the medium.

The difference between visceral and skeletal and cardiac-muscle low-ionic-strength protein extracts is also made evident when the percentages of protein which are soluble and insoluble after lowering the pH of the extract to 5·0 are compared. Of the order of fifty per cent of the protein in extracts of uterine, gastric or carotid smooth muscle is precipitated at this pH in contrast to ten per cent or less for skeletal and cardiac muscles.

Although these abherent solubility properties do probably point to some fundamental differences between the contractile proteins of smooth and striated-type muscles, it also seems likely that the large precipitates produced at a lower pH result in part from the presence of much at-present ill-defined protein occurring in smooth but absent from skeletal and cardiac muscle.

The ease of extractibility of the visceral-muscle contractile proteins is probably also related to their generally greater lability, a factor which has been particularly suspected by fine-structural workers who have until recently experienced considerable difficulty in finding fixation methods which will demonstrate the presence of the myosin component of the visceral-muscle cell. The fact that these components can be demonstrated when the correct conditions are found, does suggest a far greater sensitivity to changes in the external medium, and hence perhaps a greater solubility than is found in other types of muscle. The *in situ* stability of the actin thin filaments renders their ease of solubility less readily understandable, although it is probable that the actin is extracted as an actomyosin complex (tonoactomyosin).

The high solubility of vertebrate visceral-muscle myosin may also be due to an *in situ* labile association with a more soluble component, perhaps tropomyosin; certainly purified myosin from such a source does not display this abnormal solubility at low ionic strength.

We shall now consider in such detail as is available for visceral and other smooth muscles the individual proteins of the contractile apparatus in terms of their molecular composition and structure. Such an account is important if we are to understand the manner in which contraction takes place as a macromolecular phenomenon.

Actin

For reasons not yet wholly understood, actin appears to be a protein consistently involved in the movement of living cells. Thus, apart from its association with the contractile apparatus of muscle, it has been found in, for example, the cells of the slime moulds, in meiotically and mitotically dividing cells, sperm tails, blood platelets and plant cells. It is probable that actin is a protein of considerable antiquity, long pre-dating the other 'contractile' proteins of muscle which were added as later sophistications and refinements.

Actin occurs *in situ* in skeletal and cardiac muscle as thin filaments of quite constant diameter approximating to 5 nm. In smooth muscle diameters tend to lie between 5 and 8 nm, with a tendency to the larger diameter. Similar filaments can be prepared *in vitro* from actin extracted from muscle. These actin filaments are formed by a quaternary level of protein polymerization from globular sub-units which have been given the name G-actin (G for globular). Highly purified preparations of G-actin from most muscle sources, be they skeletal, cardiac, visceral, vertebrate or invertebrate, all appear to have very similar molecular weights. A currently accepted value is a molecular weight of $4·6 \times 10^4$. It is in this form that actin is extracted from muscle.

In terms of amino acid compositions also, most actins are very similar, although not unnaturally the further apart species are on the evolutionary ladder the greater differences in detail become. Table 7.2 compares the amino acid compositions of three types of vertebrate smooth-muscle G-actin with vertebrate skeletal and cardiac-muscle protein and an invertebrate actin. In its amino acid composition actin has nothing particular to distinguish it from most other common globular proteins, having an enzymic or biologically active function, other than the presence of 3-methyl histidine and a rather high proline content. The latter amino acid probably accounts for the low alpha helical content of the molecule

Table 7.2.—Amino acid compositions of smooth, skeletal and cardiac-muscle actins compared. Residues /60,000 g protein.

Amino acid	Smooth Muscle				Skeletal		Cardiac
	Bovine carotid	Sheep uterus	Human uterus	Mollusc (pecten)	Bovine	Rabbit	Bovine
lysine	27·0	26·7	26·8	29·0	27·6	25·2	27·3
histidine	10·9	10·5	10·1	8·3	10·7	10·1	10·2
arginine	25·6	26·0	25·0	25·4	25·1	24·4	25·3
SCM cysteine	26·4	7·3	7·0	4·2	5·1	6·7	5·9
aspartic	46·9	46·4	46·6	49·6	49·2	46·9	47·2
threonine	33·7	36·4	37·2	38·1	41·2	37·8	37·0
serine	34·0	37·7	35·9	37·2	35·6	32·3	34·6
glutamic	59·8	55·9	56·1	67·9	55·6	54·7	56·2
proline	26·2	25·9	25·8	19·3	24·4	25·2	25·2
glycine	39·7	39·8	38·9	35·7	40·2	37·4	37·5
alanine	41·4	41·9	42·1	45·4	43·0	40·5	40·7
valine	26·1	25·8	25·4	25·0	26·2	24·9	27·3
methionine	19·0	15·3	16·1	15·7	13·8	21·8	20·4
isoleucine	39·8	37·6	36·2	31·8	37·9	36·7	34·6
leucine	38·7	38·6	38·3	39·6	37·5	34·5	36·7
tyrosine	20·0	19·6	18·8	16·8	19·5	21·5	19·9
phenylalanine	16·4	16·9	16·5	15·7	16·9	15·7	16·1
tryptophane	5·0	5·4	4·5	4·3	5·0	5·1	4·7
3 methylhistidine	1·2	—	—	—	—	1·3	—
Total		515·0	507·3	509·0	514	502·7	507·0

Amino acid analyses from Gosselin-Rey *et al.* (1969), Carsten and Katz (1964), Carsten (1965).

(15–20%). G-actin has a low charge and consists of a single polypeptide chain with no stabilizing disulphide bridges. Peptide maps of skeletal and smooth muscle actins are very similar.

Hydrodynamic methods indicate that G-actin is a sphere of 6·2 nm diameter. The value obtained by electron microscopy is of the order 5·5 nm, a difference accountable for by swelling and hydration. Actin is both enzymically and structurally functional, and it follows therefore that there must be several important physiological sites of interaction and binding on the molecule.

(i) Sites of interaction with neighbouring actin molecules in the polymeric form of the molecule.
(ii) Sites of specific binding with nucleotides and divalent cations.
(iii) Sites of interaction with other muscle proteins including myosin.

It has been shown using vertebrate skeletal muscle actin preparations that in the presence of ATP, KCl and Mg^{2+}, G-actin polymerizes to give long threads of fibrous polymer corresponding to the thin filaments seen in muscle *in situ*. This fibrous polymer is generally called F-actin (F for filamentous or fibrous).

The stoichiometry of the reaction is as follows:

$$n \text{ G-actin} + n \text{ ATP} \rightarrow n \text{ ADP (F-actin)} + n \text{ Pi}$$

F-actin filaments can be shown by electron microscopy to be 'identical' with the thin filaments of smooth muscle, as well as by fluorescent-antibody staining, while closely similar thin filaments can be isolated intact from muscle by gentle homogenization in Mg^{2+} ATP and EGTA (ethyleneglycol-bis (amino ethyl ether) N.N' tetraacetic acid).

Depolymerization of the *in situ* F-actin filaments of muscle takes place completely under the extraction conditions normally applied to skeletal muscle. However, actin preparations from visceral muscle contain another protein, tropomyosin B, in greater proportions than are present in skeletal muscle. This 'contaminant' can be removed by ammonium sulphate fractionation, but even after such purification differences in the properties of some visceral-muscle depolymerized actin and skeletal-muscle actins emerge. Thus, that of the cow carotid has a much higher sedimentation rate than uterine muscle or striated-muscle actin, and it would seem that depolymerization does not always take place to the G-actin monomeric state as it does consistently for skeletal-muscle protein. The size of the aggregates appears to depend upon the preparative procedure. Such incompletely depolymerized visceral-muscle G-actin aggregates can polymerize and combine with myosin and can activate the Mg^{2+} ATPase to the same extent as striated muscle F-actin. Differences in molecular structure to account for this are not obvious.

Electron microscopy of F-actin filaments reveals that they consist of double helices made up from chains of G-actin sub-units (figure 7.1). The strands cross over every 36–37 nm along the actin filament, and the number of monomers per turn is 13–14. The helix is therefore non-integral. All this is confirmed by X-ray diffraction data. Both chains run in the same direction and have polarity as can be shown by the manner in which they bind myosin heads.

Although relatively little of the work which has established the structure of the actin thin filament has been carried out using visceral-muscle protein, it is unlikely that further work will establish significant differences between actins from different muscle sources. Some points do, however, require clarification; the sluggish depolymerization properties of some visceral-muscle actins already mentioned, the tendency of some invertebrate actins of both striated and smooth-muscle origin to have filament

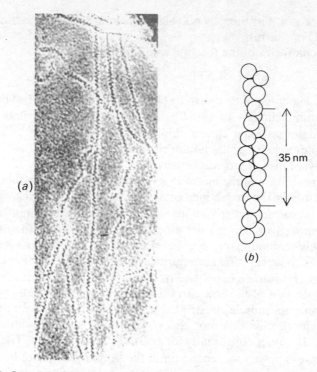

Figure 7.1. Structure of F-actin. (*a*) Electron micrograph of F-actin negatively stained with uranyl acetate. The beaded appearance of the filaments can be interpreted as shown in (*b*) as a non-integral double helix formed by the twisting of two filaments of globular units around one another. There are 13–14 sub-units per turn of helix. (Hanson and Lowy, 1963).

diameters significantly in excess of 56 nm, and the unusual appearance of some actins in invertebrate smooth muscle *in situ* when sectioned in transverse section where tubular, horseshoe and figure-of-eight profiles are seen.

Tropomyosin B and troponin in the thin filament

Rather than move directly on to consider the other quantitatively major protein of the contractile apparatus, myosin, it is appropriate at this point to discuss two other proteins which are associated with actin in the *in situ* filament.

The contraction relaxation cycle of muscle under physiological conditions is regulated by the intracellular concentration of free calcium ion.

This concentration is controlled by the sarcoplasmic reticulum of the muscle cell, but the actual sensitivity of the contractile system to calcium involves the selective interaction of a specific protein system with the calcium and with the actin-myosin interaction. This specific protein system consists of two proteins, tropomyosin and troponin, which are bound to F-actin in the thin filaments of muscle. It would seem that, when calcium is not present, troponin and tropomyosin act together to inhibit the interaction of F-actin and myosin. The binding of calcium to troponin cancels this effect. It is likely that a complex association of charge and conformational change effects are involved in the process.

Tropomyosin is the structural link which acts to position the troponin molecule along the thin filament. Ebashi and his co-workers (1969) have proposed a model for the thin filament in which the tropomyosin molecules are located in the grooves of the F-actin double helix; the effect is analogous to the locating of histones in the grooves of the DNA double helix. One tropomyosin molecule covers seven G-actin monomers and carries one troponin molecule. The repeat along the thin filaments is 40 nm, a fact neatly obtained by combining an electron-microscopic technique with labelling of the thin filaments by troponin antibody. This picture is supported by the observations that *in vitro* tropomyosin binds to F-actin, and troponin binds to tropomyosin but not to F-actin; that tropomyosin molecules aggregate end to end *in vitro*, and the shape of the molecule in crystalline tropomyosin is a curve capable of following the helical groove of F-actin; that optical reconstruction in three dimensions from thin-filament electron micrographs shows material lying between the two actin strands; that troponin-tropomyosin crystals prepared *in vitro* contain the troponin at 40 nm intervals along the tropomyosin fibrils, and that X-ray diffraction patterns of intact muscle show features consistent with a 40 nm periodicity along the thin filaments (Hanson *et al.*, 1973). Much of this work has been carried out using skeletal muscle, and details of the situation in visceral muscle are lacking.

The molecule of tropomyosin B is a completely helical two-stranded coiled-coil of M.Wt. $6.3–6.8 \times 10^4$. The presence of tropomyosin in visceral muscle has been established for some time. Early workers isolated it from human, cow and rabbit uterus, from cow bladder and from duck gizzard, as well as from cow carotid. Tropomyosins from all these sources appeared to have very similar properties with regard to their electrophoretic mobilities (Hamoir, 1973). Carsten (1968) however has compared tropomyosins from human and sheep uteri with skeletal tropomyosins. The amino acid composition (Table 7.3) of the visceral-muscle proteins

Table 7.3.—Amino acid compositions of smooth and skeletal-muscle tropomyosins (B). Residues /55,000 g protein.

Amino acid	Smooth		Skeletal	
	Human uterus	Sheep uterus	Human	Sheep
lysine	48·3	52·8	58·4	57·2
histidine	4·9	5·4	3·1	2·9
arginine	26·9	25·7	22·6	23·0
SCM cysteine	1·3	2·9	1·5	3·5
aspartic	39·1	41·8	44·4	43·0
threonine	13·5	13·8	11·7	12·1
serine	18·4	17·7	20·2	17·9
glutamic	114·1	112·9	106·0	110·0
proline	0·8	0·3	1·0	2·3
glycine	7·0	5·3	7·8	7·8
alanine	51·2	52·1	51·7	51·4
valine	14·9	15·5	13·2	12·9
methionine	10·0	10·0	10·1	8·2
isoleucine	12·0	12·1	14·4	15·1
leucine	48·3	47·7	44·7	43·6
tyrosine	5·7	5·0	7·4	7·0
phenylalanine	2·2	2·0	2·5	3·0
tryptophane	0·0	0·0	0·0	0·0
Total	418·6	423·4	420·7	420·9

Carsten (1968).

are distinctly different from the skeletal-muscle proteins, and this is borne out by the peptide maps of the proteins which showed that the tropomyosins of visceral muscle gave rise to six to eight less peptides than the skeletal proteins. Clearly there are significant differences in the primary sequences of the polypeptide chains.

As already mentioned, visceral muscle contains more extractable tropomyosin than skeletal muscle; the net content is also generally higher. Hamoir (1973), for example, suggests that cow carotid contains twice as much tropomyosin as skeletal muscle. The reason for this, as we shall see later, is that it appears to have an important role also in the organization of visceral and other smooth-muscle thick filaments.

Although troponin was originally thought to be a single homogeneous protein species, this is now known not to be the case. Skeletal-muscle troponin consists of three sub-units which have been designated TN-I, TN-T and TN-C. TN-I has a molecular weight of 24,000 and inhibits actomyosin ATPase activity, TN-T has a molecular weight of 37,000 and interacts strongly with tropomyosin, while TN-C with a molecular weight of 20,000 binds calcium. Since cardiac troponin contains sub-units of quite different molecular weight from those for the skeletal-muscle protein, comparative data for visceral muscle are awaited with interest.

Isolation of a pure preparation of troponin from visceral muscle has only recently been achieved (Carsten, 1971) although protein preparations having properties strongly favouring the identification of troponin as a visceral-muscle component were studied at a much earlier date. Pure troponin from cow uterine smooth muscle has a slightly greater gel-electrophoretic mobility than cow skeletal-muscle troponin, has inter-active properties with uterine-muscle tropomyosin, and inhibits both skeletal and visceral-muscle actomyosin ATPase activities. Hybrid complexes of skeletal-muscle tropomyosin with visceral-muscle troponin still show actomyosin ATPase inhibitory activity, albeit at a lower level than the skeletal-skeletal or smooth-smooth complexes.

Certain types of somatic and visceral smooth muscle of invertebrate origin may contain no troponin. Thus studies of the anterior byssus re-tractor muscle of bivalve molluscs indicated that the thin filaments do not contain troponin, and that calcium sensitivity and the control of activity re-side in the myosin-containing thick filaments (Kendrick-Jones et al., 1970).

Myosin—structure of the monomer

The protein myosin is with actin the other major protein component of the contractile apparatus of muscle, be it skeletal, cardiac or smooth. Like actin it exists as a monomeric molecule which in situ is normally aggregated to a greater or lesser extent to give a polymeric quaternary organized filamentous structure.

In striated muscles the actin and myosin filaments are present as separate clearly distinguishable overlapping structures. The myosin molecules are so aggregated that the polymeric myosin filaments have the globular regions of the myosin monomer projecting from the surface of the cylindrical filament with an axial periodicity of 14·3 nm (figure 7.2). These projections contain the myosin molecule ATPase activity and actin binding properties and act as cross-bridges to the actin filaments. The myosin filaments in skeletal muscle usually present an effectively rounded (if angular) profile in transverse section and, since their diameter is much greater than that of the thin actin filaments (15 nm approximately), they are generally referred to as 'thick filaments'.

In the skeletal-muscle thick and thin filaments, both the actin molecules and the myosin molecules are arranged so that there is structural polarity. Thus the projecting globular regions of the myosin molecules

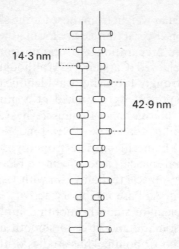

14·3 nm

42·9 nm

Figure 7.2. Schematic diagram of the arrangement of cross-bridges on a skeletal-muscle myosin-thick filament. The helix organization of the cross-bridges is 6/2e, the helical repeat 42·9 nm (Huxley, 1971).

on the myosin thick filaments do not stick straight out at right angles to the long axis of the filament, but make an acute angle with the filament in such a manner that these globular regions, cross-bridges or 'myosin-heads' point in opposite directions on either side of a 'headless' zone centred on the mid-point of the filament. Similarly, the two strands of the actin filament run in the same direction, but point in opposite directions on either side of the Z-disc to which the actin filaments attach. This polarity in the contractile protein filament system provides a basis for the sliding-filament mechanism of contraction in skeletal muscle (Chapter 8).

Although progress in the study of the contractile apparatus of smooth muscle has been less rapid than in the field of striated muscle, evidence available now suggests that the organization of myosin may not be so very different and that, as we shall see in the chapter that follows, the sliding-filament mechanism operates there also. The difficulty about vertebrate visceral muscle is that until very recently the question of the mode of organization of the contractile proteins, and in particular the myosin elements, has been vexed and controversial. More precisely, the problem has been to decide whether smooth muscle *in situ* contains myosin in a dispersed state, or as thick filaments, or whether thick filaments are only assembled at the time of contraction. Most workers have experienced difficulty in demonstrating (in the electron microscope in a consistent manner) the presence of thick filaments in smooth muscle, in spite of the chemical evidence for the presence of appropriate levels of myosin in the muscle and the physiological evidence for a sliding-filament mechanism. It now appears, however, that these failures to visualize thick filaments, and the thick filament to thin filament cross-bridges, is due to inadequate preparatory technique for electron microscopy, and that thick filaments can after all be demonstrated in relaxed, contracted, stretched and unstretched vertebrate visceral muscle (Somlyo et al., 1973).

A variety of vertebrate visceral muscles have been used as sources of myosin for physicochemical study. Myosins extracted from bovine carotid, chicken gizzard, horse oesophagus, and rabbit and human uteri, all appear to have similar size and shape properties which also resemble those of skeletal and cardiac-muscle myosins (Hamoir, 1973 review). Differences between skeletal and smooth-muscle myosins do, however, occur. Thus, if carotid and skeletal myosin from cow are compared, the smooth-muscle myosin has a higher salting-out range, a larger tendency to aggregation after reprecipitation at low ionic strength and neutral pH, a much greater thermostability in terms of precipitation when heated in solution at 100 °C, and a greater stability to tryptic digestion. The ATPase activities of visceral and skeletal-muscle myosins also differ, being much lower in smooth muscle, while their Ca-ATPase activity is higher at higher ionic strength. ATPase activity of visceral-muscle myosin plateaus when calcium concentration exceeds 5 mmol/1, while skeletal myosin continues to show an increase in activity. Visceral-muscle myosin has a low Mg-actomyosin-ATPase activity (Hamoir, 1973 review).

Now that visceral-muscle myosins have been prepared in sufficiently pure state for amino acid analysis to have some meaning, comparative studies are also possible here. The general patterns are very similar, both between different visceral-muscle preparations, and between these and skeletal-muscle myosins (Table 7.4). Significant differences do occur, and these are usually found in differences in balance between the acidic and basic amino acids of visceral and skeletal-muscle proteins, which probably result overall in visceral-muscle myosins having a slightly higher net negative charge than the skeletal proteins. Uterine smooth-muscle myosin contains 3-methyl histidine as does skeletal myosin.

Considerable effort has been put to establishing the detailed structure of the myosin molecule as it is found in striated skeletal muscle and, while less is known regarding the structure of the smooth-muscle counterpart, work is now in progress to provide this important

Table 7.4.—Amino acid compositions of smooth and skeletal-muscle myosins. Residues/10^5 g myosin

Amino acid	Smooth		Skeletal	
	Chicken gizzard	Horse oesophagus	Chicken breast	Horse
lysine	88	76	94	90
histidine	14	12	17	15
arginine	48	43	46	46
SCM cysteine or cysteine	9·9	8·4	8·3	8·2
aspartic	89	103	82	80
threonine	44	38	36	36
serine	47	33	45	36
glutamic	166	178	166	172
proline	25	17	20	22
glycine	48	28	41	37
alanine	76	77	75	75
valine	41	38	42	40
methionine	20	21	24	21
isoleucine	36	33	42	34
leucine	87	99	86	79
tyrosine	16	14	17	14
phenylalanine	27	27	30	29
tryptophane	6·6	—	6·2	—
Total	888·5	845·4	877·5	834·2

Barany *et al.* (1966). Yamaguchi *et al.* (1970).

comparative information. The skeletal-muscle myosin monomer has a molecular weight of about 500,000 and is asymmetric with a tail 140 nm long and 1·5–2·0 nm wide. The molecule has a double globular head region with approximate dimensions 15–20 nm long by 4–5 nm wide. The molecule has a complex sub-unit structure.

Trypsin dissociates skeletal-muscle myosin into two large fragments: a so-called 'light meromyosin' (LMM) of molecular weight 150,000 and a 'heavy meromyosin' (HMM) of molecular weight 340,000. HMM retains the ATPase activity of the myosin molecule. Electron-microscope studies have indicated that LMM is a rod-shaped unit 65 nm long representing the major part of the tail region of the myosin molecule, while HMM contains the globular head and a 40 nm stretch of the tail. X-ray diffraction studies have shown that LMM is a double helix of two protein chains, themselves almost entirely in the alpha helical conformation, and hence giving a very rigid rod-like structure to the myosin tail. More extensive trypsinization of HMM cleaves it further to three fragments, two of which are semi-identical and which are designated

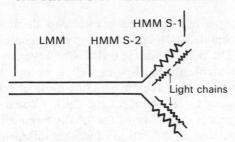

Figure 7.3. Schematic diagram of the sub-structure of the myosin monomer. LMM, light meromyosin; HMM S–2 and HMM S–1, heavy meromyosin subfragments 2 and 1 respectively.

HMM S–2 (one fragment) and HMM S–1 (two fragments). S–2 has a molecular weight of 60,000 and consists of a further rod-like section of coiled-coil double helix, i.e. it is the remainder of the myosin tail. The two S–1 fragments each have a molecular weight of 120,000 and originate one each from the paired globular heads of the intact myosin molecule. This sub-structural organization is summarized in figure 7.3. In addition to the polypeptide chains described above, the myosin molecule contains three other polypeptide chains (light chains) not peptide-bonded to the main chains. Reaction of myosin with 5, 5' dithiobis (2-nitrobenzoic acid) liberates a light-chain fraction (about half of the total light-chain content) of molecular weight 18,000 from the head region of the myosin molecule. This is achieved without loss of ATPase activity. Two more light chains are released at pH 11·4. These latter light chains called A_1 and A_2 have molecular weights of 25,000 and 16,000 respectively, and their release is accompanied by loss of myosin ATPase activity. These results obtained with intact myosin can also be obtained with trypsin fragment HMM S–1, thus placing the light chains in the head region of the molecule (figure 7.3). The 3-methyl histidine of the myosin molecule is also localized in the S–1 fragment.

Information available at present seems to suggest that in principle the sub-unit structure of the visceral-muscle myosins differs little from that of the skeletal and cardiac myosins of vertebrates.

The sub-structure of cow carotid myosin has been investigated by Huriaux and his co-workers (see Hamoir, 1973 for a review). Cow carotid LMM has been isolated and compared with skeletal-muscle myosin LMM from the same species, as well as with the skeletal protein from rabbit. The amino acid compositions of the fragments are similar, the only difference overall in the cow LMM fragments being the presence of five more residues in the skeletal myosin peptide than in the visceral-muscle myosin peptide, and a higher glutamic acid content in the visceral-muscle peptide. The visceral-muscle LMM is, however, much more soluble than its skeletal counterpart at low as well as at high ionic strength. It is believed that the LMM region contains most of the interactive information necessary for assembly of the thick filaments of the muscle, and hence the greater solubility of the LMM fragment of the visceral-muscle myosin may be an indication of weaker interactive and assembly properties of these myosins, and a reflection in the noticeably lower stability of visceral-muscle thick filaments.

Chicken gizzard myosin LMM fragment, obtained by papain digestion, has also been investigated to a limited extent. Electron microscopy reveals that this fragment is somewhat longer than the equivalent portion of the tail region from chicken skeletal-muscle myosin, the length being 156 nm as against 145 nm for the skeletal myosin.

Heavy meromyosin fractions from visceral muscle have not as yet been the subject of detailed investigation, although the S–1 fragments have been examined more closely. Cow carotid muscle S–1 fragment has a more globular shape than skeletal myosin S–1 fragments, and differs from cow cardiac myosin S–1 by having higher contents of aspartic acid and phenylalanine, which seem to be balanced by lower contents of glutamic acid and leucine. As with the skeletal-muscle peptide, the ATPase activity of the visceral-muscle myosin seems to reside in the S–1 fragment, while the characteristic low levels of ATPase activity and ionic strength sensitivity mentioned earlier as being associated with smooth-muscle myosins are also carried into the S–1 fragments.

Light-chain sub-units have been separated from chicken gizzard myosin by Kendrick-Jones (1973). The detergent sodium dodecylsulphate splits gizzard myosin into two major heavy-chain sub-units of molecular weight 200,000, and two minor components, light chains, with molecular weights of 20,000 and 17,000; these compare well with figures for skeletal myosin light chains mentioned earlier. Kendrick-Jones finds that the light chains of this smooth muscle occur in non-stoichiometric amounts in the molecule regardless of how purified it is. The two light fractions appear to be chemically quite distinct, both in amino acid composition and in tryptic peptide map, and it would seem that they are probably also distinct from the light chains of skeletal and cardiac myosins. The 17,000 molecular-weight chain appears to be involved in the hydrolytic activity of the myosin, since its removal (by urea) causes irreversible loss of ATPase activity. This is in keeping with findings regarding the role of the light chains in skeletal-muscle myosin activity. In molluscan somatic smooth muscle (catch muscle) one of the light chains appears to act as the calcium-sensitizing factor rather than, as is the case for most skeletal, cardiac or vertebrate visceral muscles, this activity being associated with the thin actin filaments.

Myosin—assembly of the thick filaments

While the problem of demonstrating myosin thick filaments in vertebrate visceral muscle appears to have been solved to the extent that these can now be seen when fixation is carefully controlled, there still remains an unresolved controversy as to whether these filaments take the form of being irregularly rounded (about 18 nm diameter) in cross-section, organized into a quasi-crystalline lattice or as ribbons. The latter case would at present seem to be more attractive, since electron-microscope evidence from intact muscle is backed up by low-angle X-ray diffraction studies and by *in vitro* studies of synthetically assembled filaments again examined in the electron microscope (Lowy *et al.*, 1973).

Working with the taenia coli muscle of guinea pig, Small and co-workers have visualized very long elements in the electron microscope and have identified them as the myosin elements by the presence on their surface of cross-bridges with a 14 nm axial periodicity (Small and Squire, 1972). These myosin elements are ribbon-shaped with a rather variable width (20–110 nm) and 10 nm thick. The cross-bridges are located on both of the ribbon faces, are directed in opposite directions on opposite faces of the ribbon, i.e. have reversed polarity, are arranged laterally, with a separation of 10 nm, in rows across the ribbon faces, while the rows of cross-bridges are separated by an axial distance of 14 nm. Low-angle X-ray diffraction studies with live relaxed and contracted taenia coli muscle (Lowy *et al.*, 1973) show as a constant feature a meridional reflection at 14 nm consistent with the existence of a regular assembly of myosin molecules in the form of filamentous elements (figure 7.4). Measurements of the shape of this reflection suggest that the diffracting elements must be at least 500 nm long in the axial direction, while diffracting coherently in the direction perpendicular to the fibre axis over a distance of at least 60 nm laterally. This latter feature can be interpreted in terms of the lateral register of the cross-bridges across the faces of the ribbon-like myosin elements seen in the electron microscope. While the intensity of the 14 nm reflection varies with the contraction state of the muscle, its shape does not. This would suggest that the

(a)

(b

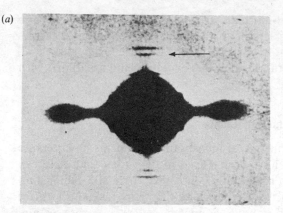

Figure 7.4. (*a*). Low-angle X-ray diffraction pattern from the resting taenia coli muscle of the guinea pig. Fibre axis vertical. The arrow indicates the 14·3 nm reflection which establishes the presence of a regular array of myosin molecules in the form of filamentous elements (Lowy *et al.*, 1973). Compare this pattern with that of a skeletal muscle (*b*) showing the clearly different pattern due to the helical arrangement of cross-bridges on the myosin filament (Huxley, 1971).

structure of the myosin elements responsible for the reflection remains unchanged throughout the contraction-relaxation cycle, and that the ribbons are present in relaxed muscles and interact with actin during contraction (Lowy *et al.*, 1973).

When low ionic strength extracts from chicken gizzard, guinea-pig taenia coli or guinea-pig vas deferens are allowed to stand in the cold for one or more days, aggregation of the proteins into two distinct types of highly-ordered filament takes place (figure 7.5) (Sobieszek and Small, 1973). The first type of filament is a long ribbon-shaped element which is up to 100 nm wide and many micrometres long. These ribbons exhibit a well-defined axial repeat period of about 5·6 nm, while optical diffraction analysis reveals a longer repeat of about 39·1 nm. The second type of filament is cylindrical, relatively short and tapered, and carries a regular arrangement of projections along their entire length, giving rise to an axial repeat of 14 nm. These latter filaments appear to be almost entirely composed of myosin and, interestingly, they seem to lack the polarity which would be necessary for participation in a sliding-filament type of contraction mechanism.

It is, however, the first type of ribbon-shaped filament which yields the most remarkable results when examined further. These ribbon-like elements appear not to be composed of myosin, but of tropomyosin, a protein which we have already discussed in its role as organizer of the

7·5(a)

7·5(b)

7·5(c)

Figure 7.5. (a). Tropomyosin ribbons from chicken gizzard smooth muscle showing the 5·6 nm periodicity. (b) Ribbon observed in the crude myosin fraction of chicken gizzard muscle, after standing 1–2 weeks, which shows a 14·0 nm repeat along part of their length. (c) Myosin filament formed from purified chicken gizzard myosin (Sobieszek and Small, 1973).

binding of troponin to the F-actin filament. Tropomyosin purified from visceral muscle can be induced to produce the same ribbon-shaped elements which form on standing the crude extracts of muscle protein in the cold. If the salt is removed from the solution containing the ribbons, they dissociate into fine filaments with an average diameter of about 8 nm, but composed of sub-filaments of 2–3 nm diameter. The 40 nm periodicity is a characteristic of tropomyosin from most muscle sources including visceral muscle.

Sobieszek and Small (1973) believe that the tropomyosin ribbons constitute a backbone structure, similar to that shown in figure 7.6, for the organization of the myosin elements in the contractile apparatus of smooth muscle. The *in vitro* aggregation of visceral-muscle myosin into cylindrical filaments, lacking polarity and without the central 'bare zone' found in self-assembled skeletal-muscle-myosin thick filaments, argues for the necessity of a core component capable of determining the polarity of the myosin molecules in the contractile elements. As already mentioned, evidence from electron microscopy and low-angle X-ray diffraction studies of smooth muscle suggests that the myosin heads, i.e. the globular regions of the myosin monomers, are arranged in an ordered manner on opposing faces of a ribbon; these 'heads' or cross-bridges have face polarity, being

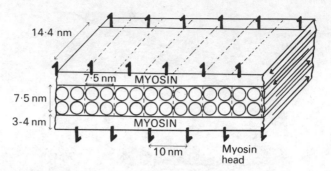

Figure 7.6. Suggested structure of a vertebrate visceral-muscle thick filament in cross-section. The backbone is composed of square-section filaments with side 7·5 nm aggregated side by side. Each square filament is composed of four cylindrical sub-units. Each side of the backbone carries a layer of myosin molecules from which myosin heads protrude (after Small and Squire, 1972).

polarized in opposite directions on opposite faces of the ribbon (Small and Squire, 1972). Visceral-muscle tropomyosin is capable of forming polar or bipolar aggregates, so there is presumably no obstacle to the formation of tropomyosin ribbons similarly polarized and capable of conferring polarity on the arrangement of myosin monomers bound to the ribbon faces. The co-operative effects of tropomyosin and myosin in assembling to produce the thick filaments observed *in vivo* is also suggested by the fact that, when the crude extracts producing ribbons and myosin filaments in the cold are allowed to stand for periods longer than a week, then many of the ribbons exhibit a rough irregular surface effectively obscuring the 5·6 nm periodicity, while in some instances this additional material imposes a new 14 nm regular axial periodicity on the ribbons, resembling the 14 nm periodicity seen on myosin ribbons in muscle in thin section. Since this periodicity on the *in situ* filaments has been identified with the regularly arranged myosin head cross-bridges, it is not unreasonable to assume that the further addition of material to the self-assembled tropomyosin filaments also represents myosin heads, and hence indicates the capability of the ribbons for assemblage and organization of the myosin monomers. Thus the finding of ribbon-shaped structures in extracts from visceral muscle which appear to be capable of interaction with myosin adds further weight to the idea that myosin-containing ribbon-shaped elements rather than cylindrical thick filaments exist *in vivo* in vertebrate visceral muscle. The major role which the tropomyosin ribbon has to play in the assembly of myosin elements in smooth muscle would, of course, account for the otherwise puzzling elevated levels of tropomyosin in smooth relative to skeletal muscle. More work will be necessary to establish whether there is a necessary incompatibility between the role of tropomyosin in the actin filament and the myosin ribbon, or whether two distinct species of tropomyosin in fact exist for the two different functions.

Squire (1971) has developed a unified theory which suggests that the way in which the molecules are packed is common to all types of myosin filament; it overcomes some of the incompatibilities of morphology between the observed structures of most skeletal-muscle thick filaments, either of vertebrate or invertebrate origin, and the filaments of visceral muscle which we have described above. Basically this theory packs the myosin monomers into a surface layer on either the cylindrical skeletal-muscle filaments or the visceral-muscle ribbons, so that the distribution of the myosin heads on the surfaces of the filaments or ribbons is the same. The myosin molecules are packed together with about 100 nm of the rod portions of the molecules associated with the backbones of the filaments or ribbons (the remainder forming the cross-bridges); the myosin rods are close packed with separations of about 2 nm, and the surface layer thickness of the packed rods is 3–4 nm. Certain implications then follow; namely, that all myosin filaments of the same backbone diameter will have approximately the same proportion of myosin to core protein by volume and weight (if the core proteins have the same density) and the cores of the filaments or ribbons will have a dimension equal to the total diameter or thickness less about 7 nm. Thus the core of the cylindrical myosin filament of vertebrate striated muscle would have a diameter of about 7 nm away from the bare zone, and with one myosin molecule per 14·4 nm repeat would have 10–15% of its protein in the core as non-myosin material. Vertebrate visceral-muscle ribbons would have a core thickness of about 7·5 nm with the core protein accounting for 25–30% of the total ribbon protein. Surprisingly, the postulated core protein of vertebrate skeletal-muscle thick filaments has yet to be isolated and characterized, or even identified with any certainty.

Paramyosin

Related in its role to visceral-muscle tropomyosin B, as a core protein organizer of myosin elements, is a protein, found in the smooth muscles of molluscs and other invertebrates, and in insect striated muscle, sometimes called tropomyosin A, but usually referred to as *paramyosin*.

Table 7.5.—Amino acid composition of invertebrate smooth-muscle paramyosins (tropomyosin A). Residues /1000 total residues.

Amino acid	Lumbricus (Annelid)	Venus (Mollusc)
lysine	67·2	72·2
histidine	12·0	4·9
arginine	103·0	99·1
cysteine	—	—
aspartic	116·0	140·0
threonine	48·0	44·1
serine	61·2	47·7
glutamic	203·0	207·0
proline	2·0	1·8
glycine	30·0	19·6
alanine	103·0	132·0
valine	34·8	34·3
methionine	1·8	13·5
isoleucine	42·0	26·9
leucine	149·0	130·0
tyrosine	12·0	22·0
phenylalanine	14·4	7·3

Kominz *et al.* (1958).

Paramyosin is a rod-shaped molecule with a fully alpha helical content arranged as a two-chain coiled coil; it thus resembles tropomyosin B and the tail of myosin. Its molecular weight is, however, greater than that of tropomyosin B: 220,000 as opposed to 68,000. In their amino acid compositions, paramyosins are noteable for having high levels of aspartic and glutamic acids, arginine, alanine and leucine (Table 7.5), and must have fairly considerable polar character. The paramyosin monomer is about 135 nm long and aggregates *in situ* together with myosin to give large filaments with a 72·5 nm axial periodicity and a prominent 14·5 nm repeat. Large-scale transverse ordering of the filaments is also recognized. Studies of paramyosin *in vitro*, using aggregates precipitated with divalent cations and examined in the electron microscope, have led to the prediction of bipolarity in the native paramyosin filament from the tendency to form both polar and bipolar arrays.

Myosin can be selectively extracted from the thick filaments of molluscan smooth muscle (catch muscle) without the solubilization of paramyosin (Szent-Gyorgyi *et al.*, 1971). Extraction of paramyosin, however, always results in solubilization of myosin. Removal of myosin changes the surface appearance of the thick filaments, producing a characteristic pattern of dark nodes or gaps in negatively stained preparations. The nodes are triangular in shape and define the polarity of the structure. Examined in this context, filaments can be found which show reversal of polarity along their length. These results have led Szent-Gyorgyi to conclude that paramyosin forms a bipolar core to the thick filament, which is covered by and imparts polarity to a layer of myosin. It is of interest to note that the molluscan myosin from this source will aggregate in the absence of paramyosin to form filaments resembling those of vertebrate skeletal muscle myosin *in vitro*. Its assembly in the molluscan thick filament is therefore a function of the paramyosin core.

Intermediate filaments

In Chapter 1 we commented upon the occurrence of a third type of filament, 10 nm in diameter, intermediate in diameter between the thin actin filaments and the thick myosin filaments. These filaments have a less electron-opaque central region and seem to be associated with the dense bodies characteristic of visceral muscle (Rice and Brady, 1973). Dense bodies and intermediate filaments can be isolated from chicken gizzard smooth muscle after exhaustive extraction of other proteins with 0·6M KCl, followed by discontinuous sucrose gradient centrifugation (Rice and Brady, 1973). Treatment with 2mM Tris over an extended period helps to separate dense bodies and intermediate filaments, and these latter can be further purified following ATP treatment and further differential centrifugation. Preliminary studies of these purified intermediate filaments indicate that they contain a protein of molecular weight about 81,000 and with an amino acid composition quite distinct from the other proteins of the muscle (Somlyo *et al.*, 1973). Intermediate filaments associated with dense bodies may function in a manner analogous to the cytoskeleton of the obliquely striated muscles of invertebrates. Clearly much more work is necessary to establish the organization of these intermediate filaments, their composition, and the constitution of the dense bodies.

Summary

It would seem that in principle most of the proteins present in the contractile apparatus of vertebrate skeletal muscles are also present in vertebrate visceral muscle, although these proteins may differ individually in details of primary, secondary and tertiary structure, and more radically in the details of their quaternary organization. In this latter respect the most profound differences would seem to be in the organization of the visceral-muscle thick filaments with their ribbon-like organization based on a core of tropomyosin carrying a polarized arrangement of myosin elements on its opposite faces. Even so this ribbon mode of organization would appear to be relateable by certain unifying principles to the cylindrical myosin filaments of skeletal muscle. The paramyosin-based myosin filaments of molluscan muscle are probably also subject to these basic rules of organization, although calcium sensitivity may reside in these latter filaments rather than in the actin filaments. Obviously many details of molecular structure remain to be clarified and much further comparative data obtained. Even so, there is probably no reason to believe, as we shall see in the next chapter, that the basic sliding-filament theories of contraction cannot be extrapolated to vertebrate visceral and other non-striated muscles, although differences in detail will undoubtedly be brought to light.

CHAPTER 8

THE MOLECULAR BASIS OF CONTRACTION IN VISCERAL MUSCLE

Most of our understanding of the way muscle works at the molecular level has been derived from studies of skeletal or striated muscle, and it is only now that knowledge of this system has achieved a relatively advanced stage that attention is being directed towards other muscle types. It is inevitable therefore in this chapter, where we consider the molecular events which occur during the contraction of visceral and other non-striated muscles, that a great deal of what will be said will be in the nature of theory and hypothesis extrapolated from the current state of understanding of striated muscle. It is appropriate, therefore, that we should begin with an account of the role of the macromolecules described in Chapter 7 in the contractile activity of skeletal muscle.

Skeletal muscle

Skeletal muscle of vertebrates is made up of cells or muscle fibres which most usually take the form of highly elongated cylinders with diameters ranging from about 50 to several hundred μm and several centimetres long. Examined under a light microscope, with polarization optics, muscle cells appear to have regular dark and light cross-striations running along the length of the cell. The dark band, which is anisotropic, is called the A-band, while the light band, which is isotropic, is called the I-band (figure 8.1). The I-bands are divided symmetrically by a thin dark line called the Z-line, while the A-bands have a centrally placed lighter zone which is called the H-zone. The H-zone itself is divided by a faint diffuse line called the M-line. The width of the A-band is constant (or usually so) but the width of the I-band changes with the state of contraction of the muscle, so that it is very much narrower in contracted muscle than it is in relaxed muscle (figure 8.1). The region between two Z-lines is called a *sarcomere*.

153

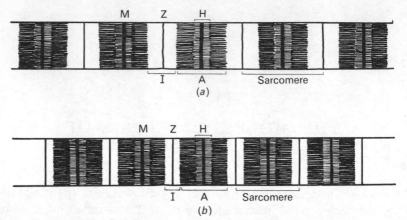

Figure 8.1. Microscopic appearance of a myofibril of skeletal muscle in longitudinal section: (a) relaxed and (b) contracted. The myofibrils seem to be made up of regularly-repeating groups of striations called *sarcomeres*. Two types of major band are seen; dark A-bands and light I-bands. The latter bands are divided by a dark line called the Z-line. For actual functional reasons (see the text) the two Z-lines on either side of an A-band are used descriptively to delineate a single sarcomere. The A-band is divided also by a lighter zone, the H-zone, and this in its turn can often be seen to be divided by a dark M-line. Note that during contraction, the A-band remains constant in length but the I-bands narrow.

The banded appearance of the skeletal muscle of vertebrates and the majority of arthropod muscles had led to the term *striated muscle* being commonly applied to this type of tissue. The striations seen in the light microscope originate in the differing refractive indices of the fibrous protein systems which are present in the cytoplasm of the cell, and which run parallel to the long axis of the cell arranged in bundles called *myofibrils*. The myofibrils contain the contractile proteins which we have discussed in Chapter 7.

Longitudinal sections of striated muscle in the electron microscope (figure 8.2) reveal thick (myosin) filaments and thin (actin) filaments lying almost exactly parallel to each other and arranged with a precision which markedly contrasts with that seen in visceral muscle (cf. figure 1.20). The thick myosin filaments, which are usually 10–15 nm in diameter, lie at equal distances from each other and are placed longitudinally exactly in register, so that all their ends are in line. Exactly half-way along each thick filament, at its central point, there is slight bulbous thickening. Each thick filament carries short projections arranged in opposing pairs at intervals of 14 nm along the filament, giving it a rough appearance at intermediate magnifications. There is, however, a zone on either side

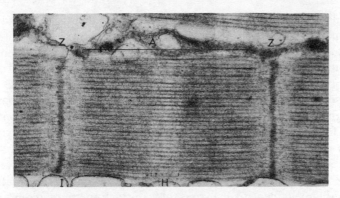

Figure 8.2. Electron micrograph of arthropod striated muscle in longitudinal section. The band lettering corresponds to that used in figure 8.1. Note how thin filaments run out from the two Z-lines (Z-discs) to overlap the thick filaments of the A-band. The M-line is not visible on this micrograph.

of the central thickening which lacks projections. The projections are the heads of the myosin molecules (Chapter 7).

Overlapping the thick filaments, but not in register with them, are the thin filaments. Like the thick filaments, the thin filaments are all also parallel to each other and exactly in register with each other. The degree of overlap of the thin filaments with the thick filaments depends upon the state of contraction of the muscle. The overlap is greatest in contracted muscle. Each thick filament is overlapped at either end by a group of thin filaments. These two groups of thin filaments extend to and connect with a band of dark material which runs transversely across the myofibril perpendicular to the filaments (figure 8.2). This dark band of material thus defines on either side a region of thick and thin filaments which is centrosymmetrical and which continues to repeat along the length of the muscle fibre. It should be clear from what we have said that this transverse dark band corresponds to the Z-line of the light micrograph, and that the region between two dark bands is the sarcomere. The thick filaments then constitute the A-band with their central thickening forming the M-line. The I-band is the thin filament region between the Z-line and the A-band of thick filaments, while the H-zone is the area where thick and thin filaments do not overlap.

During contraction, the thin filaments slide past the thick filaments, so that the Z-line (three-dimensionally it is in fact a disc) on either side makes a closer approach to the ends of the thick filaments. The I-band therefore tends to narrow during contraction. Similarly with the free ends of the thin filaments approaching each other from either end of the sarcomere during contraction, the H-zone also narrows and can disappear. In fact, some muscles are known where the apposing thin filaments have the potential to slide right past each other until they come up against the Z-line or disc at the far end of the sarcomere and, because the Z-discs in these muscles are perforated, the actin filament ends may pass right through into the next sarcomere. Such behaviour is called *supercontraction* and is found mainly in insect intersegmental and flight muscles, but also occurs in the rapid-acting musculature of the chamaeleon tongue. Observation of the movements of thin and thick filaments relative to one another have led to an explanation of muscle contraction called the 'sliding-filament theory' which says that contraction originates in a specific spatially directed interaction between myosin thick filaments and actin thin filaments resulting in relative movement (Huxley, 1971).

In transverse section, striated muscle shows the thick and thin filaments in end-on view and, depending on the point in the sarcomere taken in the section, the thin filaments alone, the thick filaments alone, or the overlapped thick and thin filaments will be seen. Such a region of overlap is seen in figure 6.4b. A common arrangement is that which is seen in this figure with a regular hexagonal array of thick filaments, and with each thick filament surrounded by six regularly spaced thin filaments in a rosette. Other arrangements are possible, for example, the hexagonally packed thick filaments of crayfish skeletal muscle have their rosettes of thin filaments in groups of 14 rather than 6, while groups of 12 are found in insect gut muscle. The projections (myosin heads) on the thick filaments form cross-bridges with the thin filaments (figure 8.2).

Any explanation of the mechanism of action of striated muscle at the molecular level has to accommodate certain observable properties of muscle. Some of these properties are physiological, some biochemical, and some ultrastructural. Among the most important of these are the observations that force developed is inversely proportional to velocity of shortening, that heat of shortening is independent of contraction speed, that energy consumption in contraction is directly proportional to work done, that contraction is accompanied by hydrolysis of adenosine triphosphate to adenosine diphosphate, that calcium moves from the sarcoplasmic reticulum into the myofibrils at contraction and back again during relaxation, and that contraction is accompanied by movement of the thin filaments past the thick filaments with a direct relationship between maximum tension developed and degree of overlap.

As we have seen in Chapter 7, a number of different proteins are involved in the organization of the myofibril, and any theory of contraction ought to be able to ascribe suitable roles for each of these proteins. It is, however, probably valid at an initial stage to consider the simplest system in which polymeric myosin (thick filaments) and polymeric actin (thin filaments) interact via cross-bridges (myosin heads) and involve in their interaction ATP and Ca^{2+}.

A very plausible theory involving these entities is that due to Davies (1963) which is summarized in figure 8.3.

The Davies theory takes as its starting point the assumption that the sliding-filament theory is the basis of contraction, and that interaction of the myosin heads on the fixed position (relatively speaking) thick filaments with the thin actin filaments causes physical translocation of the thin filaments along the line of the thick filaments. The Davies theory suggests how this movement may be brought about with large-scale contraction originating from very small-scale molecular movement, probably as conformational changes in the bulbous heads of the myosin molecules. The distances involved in these protein conformation changes must be of the order of about 10 nm, since it is possible to calculate that for every ATP hydrolysed during contraction the muscle shortens by approximately this distance.

As figure 8.3 shows, the initial state of the muscle prior to contraction is with the myosin head not in contact with the actin filament. Release of calcium from the sarcoplasmic reticulum following arrival of a nerve impulse at the cell membrane (Chapter 5) results in formation of a calcium bridge between negatively charged bound ADP on the actin filament and negatively charged bound ATP on the myosin head. Until this time the main part of the myosin head has been in an extended conformation jutting at an angle towards the calcium binding site on the actin filament, and held extended by the mutual repulsion of the negative charge on the bound ATP and another negatively charged grouping postulated to exist in a region near to the myosin head base and in the region of the myosin ATPase active centre. The formation of the calcium bridge neutralizes the negative charge on the myosin-bound ATP, and hence the effect of extension by mutual repulsion is lost. The myosin head now contracts, probably by adopting a coiled alpha helical conformation, and in so doing pulls the calcium bridge and hence the actin filament diagonally towards itself, with the net resultant of a parallel movement of the actin filament relative to the myosin filament. Adoption of a helical conformation by the myosin head has brought the ATP binding site close to the active centre of the myosin ATPase, so that hydrolysis to ADP

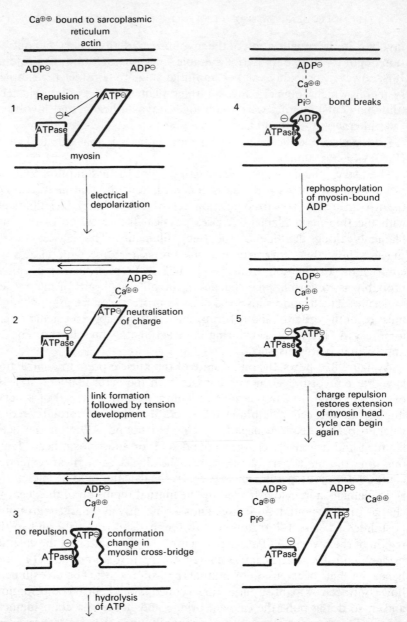

Figure 8.3. Diagrammatic representation of the events occurring during interaction of the thick-filament myosin head ATPase and ATP binding site with the ADP binding site of the actin thin filament in skeletal muscle. The effect of this series of interaction events is the sliding of the thin filament relative to the thick, and hence contraction of the muscle cell (after Davies, 1963).

and inorganic phosphate can now take place. Release of the terminal phosphate group of the ATP also breaks the calcium bridge to the actin filament, since this phosphate group was also ionically bound to the calcium. The ADP bound to the myosin head is now rephosphorylated (a number of routes are possible) restoring the negative charge on the head. Mutual repulsion between this charge and that close to the ATPase active centre can now occur, and the alpha helical conformation of the myosin head switches once again to the extended conformation. The whole cycle can begin again with the next calcium binding site along the actin filament involved this time. The cycle is envisaged as not being synchronized, so that at any one time some cross-bridges will be made and others will not. This will ensure that tension can be maintained throughout contraction. Naturally all this is hypothesis, and many of the details of conformation changes and their implications may not be at all accurate. The theory does, however, fit most of the known properties of contracting skeletal muscle.

More recently attention has switched to the role of calcium in the process of contraction; workers such as Squire have considered the possibility that the binding of calcium to troponin in the troponin-tropomyosin-actin (Chapter 7) complex causes a tropomyosin-actin conformational change which releases the inhibition of the actomyosin interaction, thereby permitting contraction to take place. Evidence for such an event comes from consideration of intensity changes in certain of the reflections in the low-angle X-ray diffraction patterns of vertebrate skeletal muscle following contraction.

Figure 8.4 shows in cross-section the relative positions to each other in a thin actin filament of actin monomers, troponin, tropomyosin, and the myosin-head cross-bridge. Electron-microscopical examination of thin filaments labelled with heavy meromyosin S–1 fragment (Chapter 7) suggests that the myosin head makes an angle of about 30° with the transverse axis of the actin double helix. In relaxed muscles the orientation of the tropomyosin molecules is probably about 45–50°, placing this close to the myosin binding site on the actin filament. Consideration of diffraction data shows that in contracting muscle the orientation of the tropomyosin changes to 65–70°, placing it much further away from the binding site of the myosin. Thus in relaxed muscle the tropomyosin is probably capable of either physically blocking the myosin binding site on the actin filament, and hence inhibiting the actomyosin interaction, or else achieves the same final effect by being so close to the binding site that it modifies the local geometry of the actin monomer involved and

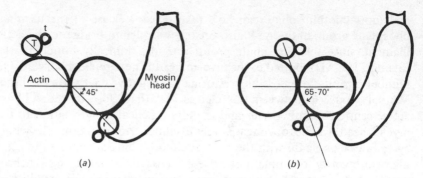

Figure 8.4. Postulated relative positions of actin molecules, tropomyosin molecules, troponin molecules and myosin heads at the point of interaction between a thin and thick filament in skeletal muscle (T, tropomyosin; t, troponin). (a) Relaxed state and (b) contracted state (after Parry and Squire, 1973). In (a) the myosin head is shown in the position it would occupy were the tropomyosin-troponin complex not in a blocking position.

thus prevents myosin attachment. Binding of calcium to the troponin molecules is likely to effect the conformation of the tropomyosin molecule in such a way as to cause a move to a position away from the myosin binding site, i.e. at 65–70°. Thus calcium binding can unlock the myosin binding site on the actin filament by modifying the conformation of tropomyosin via its bound troponin. Since tropomyosin is almost entirely helical (Chapter 7) it seems likely that the conformation change could involve some sort of supercoiling. Because of the manner in which the tropomyosin molecules are arranged along the groove of the F-actin double helix, it is also likely that the conformational change would, in transmitting itself along the full length of each tropomyosin, affect the myosin binding sites on each of its associated seven actin monomers (Parry and Squire, 1973). Further consideration of X-ray diffraction data before and after contraction provides evidence also for actual physical existence of myosin-actin cross-bridges in contracted skeletal muscle. Intensity increases of the order of 20% for diffraction layer lines at 5·1 and 5·9 nm, corresponding to actin monomer diameters and polymeric actin helix pitch respectively, can only be satisfactorily accounted for in quantitative analysis by increases in electron density at the actin molecules due to addition of extra material as myosin cross-bridges. Apart from electron-microscopical evidence which seems to indicate attachment of myosin heads to thin filaments during contraction, these diffraction data are the only experimental evidence for protein-protein cross-bridge formation in contracting muscle. This theory of cross-bridge formation regulation by tropomyosin envisages a rather different role for calcium than does the Davies hypothesis; however, this does not necessarily affect the main idea of the hypothesis that movement results from cyclical conformation change in the head region of the myosin molecule. All it modifies is the concept of the nature of the myosin-actin binding site.

Visceral muscle

Physiologically, the resemblance of smooth to striated muscle in terms of length/tension and force/velocity curves, and the identification ultrastructurally of thin actin filaments and thick myosin filaments with crossbridges, suggests a sliding-filament mechanism of operation similar to that of skeletal and cardiac muscle. This is also borne out by X-ray diffraction studies of relaxed and contracted visceral muscles which show no changes in main axial periods of actin and myosin in the transition. If this is the case, then it is obvious that some of the details of the mechanism must be different. A particular point which must be taken into account is the different mode of organization of the myosin cross-bridges of the visceralmuscle thick filaments. As we have seen in Chapter 7, the myosin thick filaments are organized so that the cross-bridges or myosin heads project in opposite directions at either end of the thick filament, thereby imparting opposite polarities, and hence permitting an interaction which draws together actin filaments from opposite ends of the sarcomere. There are no true sarcomeres in smooth muscle, and the thin filaments seem to be quite randomly distributed in the direction parallel to the long axis of the cell and the thick filaments. The polarity of the cross-bridges on the ribbon thick filaments is, however, differently arranged (Chapter 7) with the cross-bridges pointing in opposite directions, this time on opposite faces of the ribbons (thick filaments) rather than at opposite ends. Thus, a thin filament or group of thin filaments lying over one face of a myosin-faced ribbon will be moved in the opposite direction to thin filaments lying beneath the ribbon.

A model for the contractile apparatus of visceral muscle has been suggested by Small and Squire (1972) which depends partly upon the factors discussed above, and also upon observations that vertebrate visceral muscles can operate and develop a high level of tension over a greater range of lengths than vertebrate striated muscle, and readily extend up to 4 or 5 times their shortest lengths. The model also depends upon the rather controversial suggestion that in vertebrate visceral muscle the so-called dense bodies have no reality (Chapter 1), being merely artifacts of preparation originating in denatured myosin. If this were so, it would mean that the thin filaments would lack interlinking structure analogues to the Z-discs of striated muscles. Accordingly, this model confines itself to the possibility that there are many small contractile units in the cell which attach at their ends to the cell membrane (figure 8.5).

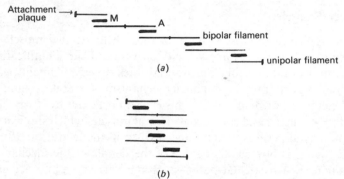

(a)

(b)

Figure 8.5. One possible arrangement of thick (M) and thin (A) filaments in a contractile unit (pseudosarcomere) of a visceral-muscle cell. The terminal thin filaments are attached to the cell membrane, are half the length of the other thin filaments, and have polarity in one direction only. (a) relaxed and (b) contracted (after Small and Squire, 1972).

Although Small and Squire give a rigorous quantitative treatment to their model, showing how its parameters can be adjusted so that a combination is found which in fact fits observed experimental data very well, it is not necessary to consider this in detail here. The model is essentially simple in principle, and can be subdivided into two cases. Basically the cross-bridges on each face of a myosin ribbon interact with a row of thin actin filaments running parallel to that face and spaced laterally at distances of about 10 nm. Although electron-microscopical evidence is lacking, it is necessary to assume that the thin filaments change polarity about halfway along their length, even though the dense bodies are assumed to be artifacts, and not sites of polarity change as are the Z-discs of striated muscle. The terminal rows of thin filaments, which are attached to the cell membrane, can be considered as being the same length as the other rows (case A) or half that length (case B). For case A a contractile unit made up of actin filaments 32 μm long and myosin ribbons 8 μm long containing 4 ribbons fits quite well with experimentally observed length tension data, while for case B the corresponding values are 36 μm, 6 μm and 5 ribbons. Contractile units are assumed to be actually half to two-thirds the total length of the cell, and staggered through the cell in an arrangement similar to that shown in figure 8.6, thereby producing tapered ends to the cell and with the force transmitted evenly over all the cell membrane.

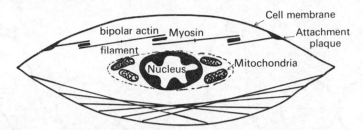

Figure 8.6. Diagrammatic representation of the possible distribution of contractile units in a visceral-muscle cell. The units are of the type shown in figure 8.5. One side of the cell shows a single unit, the other a group staggered along the cell length. In reality the scale of the cell would cause all such units to lie very nearly parallel to each other (modified after Small and Squire, 1972).

The situation regarding regulation of contraction by calcium in smooth muscle is somewhat clouded. Both molluscan and vertebrate somatic and visceral smooth muscles have been studied by Parry and Squire (1973) by X-ray diffraction methods in a manner similar to that described above for skeletal striated muscle. More definite results were obtained using anterior byssus retractor muscle from *Mytilus* as a source of diffraction data than vertebrate taenia coli, and we shall therefore comment on the former muscle first. It must, however, be remembered that byssus retractor muscle is apparently different from vertebrate smooth muscle (and vertebrate skeletal as well as arthropod muscle) in that it has its calcium-sensitive properties associated with the myosin thick filaments, that the presence of troponin is uncertain (but probably absent as the calcium binding component), and perhaps less importantly that the myosin ribbon has a backbone of paramyosin rather than tropomyosin B. We can ignore arguments about the functioning of this muscle as a catch muscle (a muscle in which the decay of tonic contraction is very very slow) since whether the muscle is behaving as a whole tonically or phasically is merely an effect of differences of neurochemical control, and irrelevant to the manner in which the actin and myosin interact. Much confusion seems to exist in the minds of some molecular biologists about this latter point. It should be quite clear from what has passed earlier that the state of the muscle depends upon whether calcium is made available or withdrawn, and that this is under the control of information arriving at the muscle-cell membrane. Hence a state of semi-permanent contraction can be maintained, without the requirement of any different architecture in the contractile apparatus, purely by neurochemical control of membrane permeability.

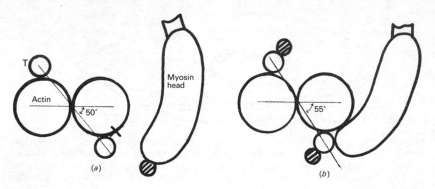

Figure 8.7. The analogous situation to that seen in figure 8.4 for a molluscan smooth muscle. Note that here in the relaxed state the calcium sensitivity resides at the myosin rather than the actin. (*a*) Relaxed state and (*b*) contracted state. The myosin head is postulated as being initially some distance from its binding site, which is marked with a cross in (*a*). The calcium-sensitive factor is cross-hatched (adapted and redrawn from Parry and Squire, 1973).

X-ray diffraction studies do show that there are changes in thin-filament structure in molluscan smooth muscle during contraction. Since there is apparently no calcium-sensitive protein permanently bound to the thin filaments, this may imply that the changes result either from association of myosin with actin, or a transfer of the calcium-sensitive factor to the thin from the thick filaments during the period of activation. Model-building studies indicate that the changes in diffraction intensities produced during activation could feasibly originate in a 5° shift of the tropomyosin molecules from their original positions occupied on the actin filaments in the relaxed state of the muscle (figure 8.7). Thus the proximity of the tropomyosin to the actin-myosin interaction site is still close. Clearly the shift and the degree of exposure of the site is much less than is suggested to take place in striated muscle, but this may be due to the different geometry involved in the association of rows of thin filaments with myosin heads protruding from a flat ribbon face, and the association of orbits of thin filaments with myosin heads arranged on a cylindrical surface. In the scheme shown in figure 8.7 the calcium-sensitive factor (presumed to have real molecular existence) 'flip-flops' to and fro between myosin and actin-tropomyosin during activation, contraction and relaxation. This case seems likely, since something has to induce the observed positional movements of the tropomyosins on the actin filament.

X-ray diffraction studies of vertebrate visceral muscle are complicated by the fact that the diffraction pattern sharpens considerably during activation, and there is obviously a great increase in molecular order in the myofilament system. However, the intensity changes in the diffraction maxima seem to suggest that the situation in this muscle is very similar to that in molluscan muscle, the movements of the tropomyosin molecules being very small.

FURTHER READING (key reviews)

Chapter 1
Burnstock, G. (1970) 'Structure of Smooth Muscle and its Innervation', *Smooth Muscle*, pp. 1–69, Ed. Bulbring, E., Brading, A. F., Jones, A. W. and Tomita, T., Edward Arnold, London.
Campbell, G. and Burnstock, G. (1968) 'Comparative Physiology of Gastrointestinal Motility', *Handbook of Physiology*, Vol. 4, section 6—Alimentary Canal, pp. 2213–2266, American Physiological Society, Washington.

Chapter 2
Wyburn, G. M. (1960) *The Nervous System*, Academic Press, London.
Campbell, G. (1970) 'Autonomic Nervous Supply to Effector Tissues', *Smooth Muscle* (see Chapter 1).

Chapter 3
Burnstock, G. (1970) (see Chapter 1).
Daniel, E. E. (1973) 'A Conceptual Analysis of the Pharmacology of Gastrointestinal Motility', *Pharmacology of Gastrointestinal Motility and Secretion*, Vol. 1, Ed. Holton, P., Pergamon Press, Oxford.

Chapter 4
Burnstock, G., Holman, M. E. and Prosser, C. L. 'Electrophysiology of Smooth Muscle', *Physiol. Rev.* **43**, 482–527.
Katz, B. (1966) *Nerve, Muscle and Synapse*, McGraw-Hill, New York.

Chapter 5
Sandow, A. (1965) 'Excitation-contraction Coupling in Skeletal Muscle', *Pharmac. Rev.* **17**, 265–320.
Somlyo, A. P. (1972) 'Excitation-contraction Coupling in Vertebrate Smooth Muscle: Correlation of Ultrastructure with Function', *The Physiologist* **15**, 338–348.

Chapter 6
Florey, E. (1966) *An Introduction to General and Comparative Animal Physiology*, Saunders Co., Philadelphia.
Sandow, A. (1970) 'Skeletal Muscle', *An. Rev. Physiol.* **32**, 87–138.

Chapters 7 and 8
Taylor, E. W. (1972) 'Chemistry of Muscle Contraction', *An. Rev. Biochem.* **41**, 577–616.
Young, M. (1969) 'The Molecular Basis of Muscle Contraction', *An. Rev. Biochem.* **38**, 913–950.

COMPREHENSIVE BIBLIOGRAPHY
(including specialist papers)

Abbott, B. C. and Lowy, J. (1953) 'Mechanical Properties of *Mytilus* Muscle', *J. Physiol.* (*Lond.*) **120**, 50 P.

Abbott, B. C. and Lowy, J. (1957) 'Stress Relaxation in Various Muscles', *Proc. R. Soc. Lond. B* **146**, 280–288.

Abe, Y. and Tomita, T. (1968) 'Cable Properties of Smooth Muscle', *J. Physiol.* (*Lond.*) **196**, 87–100.

Anderson, N. C. (1969) 'Voltage Clamp Studies on Uterine Smooth Muscle', *J. gen. Physiol.* **54**, 145–165.

Anderson, N. C., Ramon, F. and Snyder, A. (1971) 'Studies on Calcium and Sodium in Uterine Smooth Muscle Excitation under Current Clamp and Voltage Clamp Conditions', *J. gen. Physiol.* **58**, 322–339.

Atwood, H. L., Luff, A. R., Morin, W. A. and Sherman, R. G. (1971) 'Dense Cored Vesicles at Neuromuscular Synapses of Arthropods and Vertebrates', *Experientia* **27**, 816–817.

Auber, J. (1967) 'Distribution of the Two Kinds of Myofilaments in Insect Muscles', *Amer. Zool.* **7**, 451–456.

Barany, M., Barany, K., Gaetjens, E. and Bailin, G. (1966) 'Chicken Gizzard Myosin', *Arch. Biochem. Biophys.* **113**, 205–221.

Barr, L. M. (1959) 'Distribution of Ions in Intestinal Smooth Muscle', *Proc. Soc. Exp. Biol. Med.* **101**, 283–285.

Baskin, R. J. (1967) 'Changes in Volume in Striated Muscle', *Amer. Zool.* **7**, 593–601.

Bell, C. (1969) 'Transmission from Vasoconstrictor and Vasodilator Nerves to Single Smooth Muscle Cells of the Guinea-pig Uterine Artery', *J. Physiol.* (*Lond.*) **205**, 695–708.

Belleau, B. (1964) 'A Molecular Theory of Drug Action based on Induced Conformational Perturbations of Receptors', *J. Med. Chem.* **7**, 776–784.

Belton, P. and Brown, B. E. (1969) 'The Electrical Activity of Cockroach Visceral Muscle Fibres', *Comp. Biochem. Physiol.* **28**, 853–863.

Bennett, M. R. (1967) 'The Effect of Cations on the Electrical Properties of the Smooth Muscle Cells of the Guinea-pig Vas Deferens', *J. Physiol.* (*Lond.*) **190**, 465–479.

Bennett, T. (1970) 'Interaction of Nerve-mediated Excitation and Inhibition of Single Smooth Muscle Cells of the Avian Gizzard', *Comp. Biochem. Physiol.* **32**, 669–680.

Bianchi, C. P. (1969) 'Pharmacology of EC Coupling in Muscle', *Fed. Proc.* **28**, 1624–1628.

Birks, R. I. and Davey, D. F. (1969) 'Osmotic Responses demonstrating the Extracellular Character of the Sarcoplasmic Reticulum', *J. Physiol.* (*Lond.*) **202**, 171–188.

Bolton, T. B. (1968) 'Electrical and Mechanical Activity of the Longitudinal Muscle of the Anterior Mesenteric Artery of the Domestic Fowl', *J. Physiol.* (*Lond.*) **196**, 283–292.

Borys, H. K. and Karler, R. (1971) 'Effects of Caffeine on the Intracellular Distribution of Calcium in Frog Sartorius Muscle', *J. Cell. Physiol.* **78**, 387–404.

Bozler, E. (1948) 'Conduction, Automaticity and Tonus of Mammalian Visceral Muscle', *Experientia* **4**, 213–218.

Bozler, E. (1965) 'Osmotic Properties of Amphibian Muscles', *J. gen. Physiol.* **49,** 37–45.

Bradbury, S. J. (1973*a*) 'The Effect of Parathion on Crustacean Skeletal Muscle. I. The Mechanical Threshold and Dependence on Ca^{2+} Ions', *Comp. Biochem. Physiol.* **44A,** 1021–1032.

Bradbury, S. J. (1973*b*) 'The Effect of Parathion on Crustacean Skeletal Muscle. II. Disruption of Excitation-contraction Coupling', *Comp. Biochem. Physiol.* **44A,** 1033–1046.

Brown, B. E. and Nagai, T. (1969) 'Insect Visceral Muscle: Neural Relations of the Proctodeal Muscles of the Cockroach', *Comp. Biochem. Physiol.* **15,** 1767–1783.

Bulbring, E. and Kuriyama, H. (1963) 'Effects of Changes in the External Sodium and Calcium Concentrations on Spontaneous Electrical Activity in Smooth Muscle of Guinea-pig Taenia Coli', *J. Physiol.* (*Lond.*) **166,** 29–58.

Bulbring, E. and Tomita, T. (1970) 'Effects of Calcium Removal on the Smooth Muscle of the Guinea-pig Taenia Coli', *J. Physiol.* (*Lond.*) **210,** 217–232.

Bullock, T. H. and Horridge, G. A. (1965) *Structure and Function in the Nervous Systems of Invertebrates*, W. H. Freeman & Co., San Francisco.

Burnstock, G. (1970) 'Structure of Smooth Muscle and its Innervation', in *Smooth Muscle*, Ed. Bulbring, E., Brading, A. F., Jones, A. W. and Tomita, T., Edward Arnold, London.

Burnstock, G. and Prosser, C. L. (1960*a*) 'Delayed Repolarization in Smooth Muscles', *Proc. Soc. Exp. Biol. Med.* **103,** 269–270.

Burnstock, G. and Prosser, C. L. (1960*b*) 'Conduction in Smooth Muscles: Comparative Electrical Properties', *Amer. J. Physiol.* **199,** 553–559.

Burnstock, G. and Wood, M. (1967) 'Innervation of the Urinary Bladder of the Sleepy Lizard (*Trachysaurus rugosus*). II. Physiology and Pharmacology', *Comp. Biochem. Physiol.* **20,** 675–690.

Burnstock, G., Greenberg, M. J., Kirby, S. and Willis, A. G. (1967) 'An Electrophysiological and Pharmacological Study of Visceral Smooth Muscle and its Innervation in a Mollusc, *Poneroplax albida*', *Comp. Biochem. Physiol.* **23,** 407–429.

Burnstock, G., Holman, M. E. and Prosser, C. L. (1963) 'Electrophysiology of Smooth Muscle', *Physiol. Rev.* **43,** 482–527.

Campbell, G. (1969) 'The Autonomic Innervation of the Stomach of a Toad (*Bufo marinus*)', *Comp. Biochem. Physiol.* **31,** 693–706.

Campbell, G. (1970) 'Autonomic Nervous Supply to Effector Tissues', in *Smooth Muscle*, Ed. Bulbring, E., Brading, A. F., Jones, A. W. and Tomita, T., Edward Arnold, London.

Campbell, G. and Burnstock, G. (1968) 'Comparative Physiology of Gastrointestinal Motility', *Handbook of Physiology*, Section 6, Vol. 4. 2213–2266, American Physiological Society, Washington.

Carroll, P. M. (1969) 'The Effect of Hypertonic Solution on the Wet Weight and Contractions of Rat Uterus and Vas Deferens', *J. gen. Physiol.* **53,** 590–607.

Carsten, M. E. (1965) 'A Study of Uterine Actin', *Biochemistry* **4,** 1049–1054.

Carsten, M. E. (1968) 'Tropomyosin from Smooth Muscle of the Uterus', *Biochemistry* **7,** 960–967.

Carsten, M. E. (1971) 'Uterine Smooth Muscle Troponin', *Arch. Biochem. Biophys.* **147,** 353–357.

Carsten, M. E. and Katz, A. M. (1964) 'Actin, a Comparative Study', *Biochim. Biophys. Acta.* **90,** 534–541.

Carter, N. W., Rector, F. C., Campion, D. S. and Seldin, D. W. (1967) 'Measurement of Intracellular pH of Skeletal Muscle with pH-sensitive Glass Microelectrodes', *J. Clin. Invest.* **46,** 920–925.

Carvalho, A. P. (1966) 'Binding of Cations by Microsomes from Rabbit Skeletal Muscle', *J. Cell. Physiol.* **67,** 73–84.

Carvalho, A. P. (1968a) 'Effects of Potentiators of Muscular Contraction on Binding of Cations by Sarcoplasmic Reticulum', *J. gen. Physiol.* **51**, 427–442.

Carvalho, A. P. (1968b) 'Calcium Binding Properties of Sarcoplasmic Reticulum as influenced by ATP, Caffeine, Quinine and Local Anesthetics', *J. gen. Physiol.* **52**, 622–642.

Carvalho, A. P. and Leo, B. (1967) 'Effects of ATP on the Interaction of Ca^{2+}, Mg^{2+} and K^+ with Fragmented Sarcoplasmic Reticulum isolated from Rabbit Skeletal Muscle', *J. gen. Physiol.* **50**, 1327–1352.

Casteels, R. (1969) 'Calculation of the Membrane Potential in Smooth Muscle Cells of the Guinea-pig's Taenia Coli by the Goldman Equation', *J. Physiol. (Lond.)* **205**, 193–208.

Casteels, R. and Kuriyama, H. (1966) 'Membrane Potential and Ion Content in the Smooth Muscle of the Guinea-pig's Taenia Coli at Different External Potassium Concentrations', *J. Physiol. (Lond.)* **184**, 120–130.

Cole, D. E. (1950) 'The Effect of Oestradiol on the Rat Uterus', *J. Endocrinol.* **7**, 12–23.

Csapo, A. (1955) *Modern Trends in Obstetrics and Gynecology*, Harper, New York.

Csapo, A. (1960) 'Molecular Structure and Function of Smooth Muscle', in *The Structure and Function of Muscle*, Ed. G. H. Bourne, Academic Press, New York.

Daniel, E. E. (1973) 'A Conceptual Analysis of the Pharmacology of Gastrointestinal Motility', in *Pharmacology of Gastrointestinal Motility and Secretion*, Vol. I, Ed. Holton, P., Pergamon Press, Oxford.

Daniel, E. E. and Singh, H. (1958) 'The Electrical Properties of the Smooth Muscle Cell Membrane', *Can. J. Biochem. Physiol.* **36**, 959–975.

Davies, R. E. (1963) 'A Molecular Theory of Muscle Contraction', *Nature* **199**, 1068–1072.

Ebashi, S., Endo, M. and Ohtsuki, I. (1969) 'Control of Muscle Contraction', *Quart. Rev. Biophys.* **2**, 351–370.

Eccles, J. C. (1965) *The Physiology of Synapses*, Springer Verlag, Berlin.

Endo, M. and Nakajima, Y. (1973) 'Release of Calcium induced by Depolarization of the Sarcoplasmic Reticular Membranes', *Nature* **246**, 216–218.

Fenn, W. O. (1936) 'The Role of Tissue Spaces in the Osmotic Equilibrium of Frog Muscles in Hypotonic and Hypertonic Solutions', *J. cell. comp. Physiol.* **9**, 93–103.

Ferry, C. B. (1967) 'The Innervation of the Vas Deferens of the Guinea-pig', *J. Physiol. (Lond.)* **192**, 463–478.

Florey, E. (1966) *An Introduction to General and Comparative Animal Physiology*, W. B. Saunders, Philadelphia.

Florey, E. and Kriebel, M. E. (1969) 'Electrical and Mechanical Responses of Chromatophore Muscle Fibres of the Squid, *Loligo opalescens*, to Nerve Stimulation and Drugs', *Z. vergl. Physiol.* **65**, 98–130.

Foulks, J. G., Pacey, J. A. and Perry, F. A. (1965) 'Contractures and Swelling of the Transverse Tubules during Chloride Withdrawal in Frog Skeletal Muscle', *J. Physiol. (Lond.)* **180**, 96–115.

Funaki, S. (1961) 'Spontaneous Spike Discharges of Vascular Smooth Muscle', *Nature* **191**, 1102–1103.

Gabella, G. (1973) 'Fine Structure of Smooth Muscle', *Phil. Trans. R. Soc. Lond. B.* **265**, 7–16.

Gillespie, J. S. (1962) 'Spontaneous Mechanical and Electrical Activity of Stretched and Unstretched Intestinal Smooth Muscle Cells and their Response to Sympathetic Nerve Stimulation', *J. Physiol. (Lond.)* **162**, 54–75.

Gonella, J. (1965) 'Variation de l'activité électrique spontanée du duodenum du lapid avec le lieu dénervation', *C. R. Acad. Sci. Paris* **260**, 5362–5365.

Goodford, P. J. (1970) 'Ionic Interactions in Smooth Muscle', in *Smooth Muscle*, Ed. Bulbring, E., Brading, A. F., Jones, A. W. and Tomita, T., Edward Arnold, London.

Gosselin-Rey, C., Gerday, C., Gaspar-Godfroid, A. and Carsten, M. E. (1969) 'Amino Acid Analysis and Peptide Mapping of Bovine Carotid Actin', *Biochim. Biophys. Acta.* **175**, 165–173.

Guth, L. (1968) 'Trophic Influence of Nerve on Muscle', *Physiol. Rev.* **48**, 645–687.

Hagopian, M. and Spiro, D. (1967) 'The Sarcoplasmic Reticulum and its Association with the T-System in an Insect', *J. Cell Biol.* **32**, 535–545.

Hamoir, G. (1973) 'Extractability and Properties of the Contractile Proteins of Vertebrate Smooth Muscle', *Phil. Trans. R. Soc. Lond. B* **265**, 169–181.

Hanson, J. and Lowy, J. (1960) 'Structure and Function of the Contractile Apparatus in the Muscles of Invertebrate Animals', in *The Structure and Function of Muscle*, Ed. G. H. Bourne, Academic Press, New York.

Hanson, J. and Lowy, J. (1963) 'Structure of F-actin and of Actin Filaments isolated from Muscle', *J. Mol. Biol.* **6**, 46–60.

Hanson, J., Lednev, V., O'Brien, E. J. and Bennet, P. M. (1973) 'Structure of the Actin-containing Filaments in Vertebrate Skeletal Muscle', *Cold Spring Harbour Symp. Quant. Biol.* **37**, 311–318.

Heistracher, P. and Hunt, C. C. (1969) 'The Relation of Membrane Changes to Contraction in Twitch Muscle Fibres', *J. Physiol. (Lond.)* **201**, 589–611.

Heuman, H. G. (1973) 'Smooth Muscle: Contraction Hypothesis based on the Arrangement of Actin and Myosin Filaments in Different States of Contraction', *Phil. Trans. R. Soc. Lond. B* **265**, 213–217.

Hill, A. V. (1938) 'The Heat of Shortening and the Dynamic Constants of Muscle', *Proc. R. Soc. Lond. B* **126**, 136–193.

Hill, A. V. (1950) 'The Development of the Active State of Muscle during the Latent Period', *Proc. R. Soc. Lond. B* **137**, 320–329.

Hill, D. K. (1968) 'Tension due to Interaction between the Sliding Filaments in Resting Striated Muscle. The Effect of Stimulation', *J. Physiol. (Lond.)* **199**, 637–684.

Hodgkin, A. L. and Horowicz, P. (1957) 'The Differential Action of Hypertonic Solutions on the Twitch and Action Potential of a Muscle Fibre', *J. Physiol. (Lond.)* **136**, 17–18 *P*.

Hodgkin, A. L. and Horowicz, P. (1960) 'Potassium Contractures in Single Muscle Fibres', *J. Physiol. (Lond.)* **153**, 384–403.

Howell, J. N. (1969) 'A Lesion of the Transverse Tubules of Skeletal Muscle', *J. Physiol. (Lond.)* **201**, 515–533.

Hoyle, G. (1961) 'Functional Contracture in a Spiracular Muscle', *J. Insect Physiol.* **7**, 305–314.

Huddart, H. (1969) 'Caffeine Activation of Crab Skeletal Muscle', *Comp. Biochem. Physiol.* **29**, 1031–1038.

Huddart, H. (1971) 'The Effect of Quinine on Tension Development, Membrane Potentials and Excitation-contraction Coupling of Crab Skeletal Muscle Fibres', *J. Physiol. (Lond.)* **216**, 641–657.

Huddart, H. (1974) *The Comparative Structure and Function of Muscle*, Pergamon Press, Oxford.

Huddart, H. and Abram, G. R. (1969) 'Modification of Excitation-contraction Coupling in Locust Skeletal Muscle induced by Caffeine', *J. Exp. Zool.* **171**, 49–58.

Huddart, H. and Bradbury, S. J. (1972) 'Fine Structure of a Neurosecretory Axon in a Crustacean Skeletal Muscle', *Experientia* **28**, 950–951.

Huddart, H. and Oates, K. (1970) 'Localization of the Intracellular Site of Action of Caffeine on Skeletal Muscle', *Comp. Biochem. Physiol.* **36**, 677–682.

Huddart, H., Greenwood, M. and Williams, A. J. (1974) 'The Effect of Some Organophosphorous and Organochlorine Compounds on Calcium Uptake by Sarcoplasmic Reticulum isolated from Insect and Crustacean Skeletal Muscle', *J. Comp. Physiol.* **93**, 139–150.

Hurwitz, L., Fitzpatrick, D. F., Debbas, G. and Landon, J. (1973) 'Localization of Calcium Pump Activity in Smooth Muscle', *Science* **179**, 384–386.

Huxley, H. E. (1964) 'Evidence for the Continuity between the Central Elements of the Triads and Extracellular Space in Frog Sartorius Muscle', *Nature* **202**, 1067–1071.

Huxley, H. E. (1971) 'The Structural Basis of Muscular Contraction', *Proc. R. Soc. Lond. B* **178**, 131–149.

Isaacson, A. (1969) 'Caffeine-induced Contractures and Related Calcium Movements in Hypertonic Media', *Experientia* **25**, 1263–1265.

Isaacson, A., Yamaji, K. and Sandow, A. (1970) Quinine Contractures and Ca^{45} Movements of Frog Sartorius Muscles as affected by pH', *J. Pharmac. Exp. Ther.* **171**, 26–31.

Ito, Y. and Kuriyama, H. (1971*a*) 'Membrane Properties of the Smooth Muscle Fibres of the Guinea-pig Portal Vein', *J. Physiol. (Lond.)* **214**, 427–441.

Ito, Y. and Kuriyama, H. (1971*b*) 'Nervous Control of the Motility of the Alimentary Canal of the Silver Carp', *J. Exp. Biol.* **55**, 469–487.

Job, D. D. (1969) 'Ionic Basis of Intestinal Electrical Activity', *Amer. J. Phys.* **217**, 1534–1541.

Johansson, B. and Ljung, B. (1967) 'Sympathetic Control of Rhythmically Active Vascular Smooth Muscle as studied by a Nerve-muscle Preparation of Portal Vein', *Acta Physiol. Scand.* **70**, 299–311.

Jonsson, O. (1969) 'Changes in the Activity of Isolated Vascular Smooth Muscle in response to Reduced Osmolarity', *Acta. Physiol. Scand.* **77**, 191–200.

Kendrick-Jones, J. (1973) 'The Sub Unit Structure of Gizzard Myosin', *Phil. Trans. R. Soc. Lond. B* **265**, 183–189.

Kendrick-Jones, J., Lehman, W. and Szent-Gyorgyi, A. G. (1970) 'Regulation in Molluscan Muscles', *J. Mol. Biol.* **54**, 313–326.

Kobayashi, M. (1969) 'Effect of Calcium on Electrical Activity in Smooth Muscle Cells of Cat Ureter', *Amer. J. Physiol.* **216**, 1279–1285.

Kobayashi, M., Prosser, C. L. and Nagai, T. (1967) 'Electrical Properties of Intestinal Muscle as measured intracellularly and extracellularly', *Amer. J. Physiol.* **213**, 275–286.

Kominz, D. R., Saad, F. and Laky, K. (1958) 'Chemical Characteristics of Annelid, Molluscan and Arthropod Tropomyosins', *Conf. Chem. Muscular Contraction*, Tokyo, 1957, 66–76.

Kuriyama, H. (1970) 'Effects of Ions and Drugs on the Electrical Activity of Smooth Muscle', in *Smooth Muscle*, Ed. Bulbring, E., Brading, A. F., Jones, A. W. and Tomita, T., Edward Arnold, London.

Kuriyama, H. and Tomita, T. (1965) 'The Responses of Single Smooth Muscle Cells of Guinea-pig Taenia Coli to Intracellularly Applied Currents, and their Effects on the Spontaneous Electrical Activity', *J. Physiol. (Lond.)* **178**, 270–289.

Kuriyama, H., Osa, T. and Toida, N. (1967*a*) 'Membrane Properties of the Smooth Muscle of the Guinea-pig Ureter', *J. Physiol. (Lond.)* **191**, 225–238.

Kuriyama, H., Osa, T. and Toida, N. (1967*b*) 'Electrophysiological Study of the Intestinal Smooth Muscle of the Guinea-pig', *J. Physiol. (Lond.)* **191**, 239–255.

Kushmerick, M. J. and Davies, R. E. (1969) 'The Chemical Energetics of Muscle Contraction. II The Chemistry, Efficiency and Power of Maximal Working Sartorius Muscles', *Proc. R. Soc. Lond. B.* **174**, 315–353.

Kushmerick, M. J., Larson, R. E. and Davies, R. E. (1969) 'The Chemical Energetics of Muscle Contraction. I. Activation Heat, Heat of Shortening and ATP Utilization for Activation-relaxation Processes', *Proc. R. Soc. Lond. B.* **174**, 293–313.

Lowy, J., Vibert, P. J. and Haselgrove, J. C. (1973) 'The Structure of the Myosin Elements in Vertebrate Smooth Muscles', *Phil. Trans. R. Soc. Lond. B.* **265**, 191–196.

McLean, J. R. and Burnstock, G. (1967) 'Innervation of the Urinary Bladder of the Sleepy Lizard (*Trachysaurus rugosus*). I. Fluorescent Histochemical Localization of Catecholamines', *Comp. Biochem. Physiol.* **20**, 667–673.

Marshall, J. M. and Miller, M. D. (1964) 'Effects of Metabolic Inhibitors on the Rat Uterus and on its response to Oxytocin', *Amer. J. Physiol.* **206**, 437–442.

Mekata, F. (1971) 'Electrophysiological Studies of the Smooth Muscle Cell Membrane of the Rabbit Common Carotid Artery', *J. gen. Physiol.* **57**, 738–751.

Michelson, M. J. (1973) 'Structure and Mutual Disposition of Cholinoreceptors and Changes in their Disposition in the Course of Evolution', in *Comparative Pharmacology*, Vol. 1, Ed. M. J. Michelson, Pergamon Press, Oxford.

Michelson, M. J. and Zeimal, E. V. (1973) *Acetylcholine*, Pergamon Press, Oxford.

Miyamoto, H. and Kasai, M. (1973) 'Re-examination of Electrical Stimulation of SR Fragments *in vitro*', *J. gen. Physiol.* **62**, 773–786.

Mommaerts, W. F. M. M., Brady, A. J. and Abbott, B. C. (1961) 'Major Problems in Muscle Physiology', *Ann. Rev. Physiol.* **23**, 529–576.

Nagai, T. and Brown, B. E. (1969) 'Insect Visceral Muscle. Electrical Potentials and Contraction in Fibres of the Cockroach Proctodeum', *J. Insect Physiol.* **15**, 2151–2167.

Nakamaru, Y. and Schwartz, A. (1972) 'The Influence of Hydrogen Ion Concentration on Calcium Binding and Release by Skeletal Muscle Sarcoplasmic Reticulum', *J. gen. Physiol.* **59**, 22–32.

Nilsson, S. and Fange, R. (1969) 'Adrenergic and Cholinergic Vagal Effects on the Stomach of a Teleost (*Gadus morhua*)', *Comp. Biochem. Physiol.* **30**, 691–694.

Osborne, M. P., Finlayson, L. H. and Rice, M. (1971) Neurosecretory Endings associated with Striated Muscles in Three Insects (*Schistocerca*, *Carausius* and *Phormia*) and a Frog (*Rana*)', *Z. Zellforsch.* **116**, 391–404.

Parry, D. A. D. and Squire, J. M. (1973) 'Structural Role of Tropomyosin in Muscle Regulation. Analysis of the X-ray Diffraction Patterns from Relaxed and Contracting Muscle', *J. Mol. Biol.* **75**, 33–55.

Pilar, G. and Vaughan, P. C. (1969) 'Electrophysiological Investigations of the Pigeon Iris Neuromuscular Junctions', *Comp. Biochem. Physiol.* **29**, 51–72.

Prosser, C. L., Nystrom, R. A. and Nagai, T. (1965) 'Electrical and Mechanical Activity in Intestinal Muscles of Several Invertebrate Animals', *Comp. Biochem. Physiol.* **14**, 53–70.

Rice, R. V. and Brady, A. C. (1973) 'Biochemical and Ultrastructural Studies on Vertebrate Smooth Muscle', *Cold Spring Harbour Symp. Quant. Biol.* **37**, 429–438.

Rushton, W. A. H. (1937) 'Initiation of the Propagative Disturbance', *Proc. R. Soc. Lond. B* **124**, 210–243.

Sakai, T., Geffner, E. S. and Sandow, A. (1970) 'Caffeine Contracture in Muscle with Disrupted Transverse Tubules', *Amer. J. Physiol.* **220**, 712–717.

Sandow, A. (1961) 'Energetics of Muscular Contraction', in *Biophysics of Physiological and Pharmacological Actions*, Ed. A. M. Shanes, Am. Ass. Adv. Sci., Washington, D.C.

Sandow, A. (1965) 'Excitation-contraction Coupling in Skeletal Muscle', *Pharmac. Rev.* **17**, 265–320.

Sandow, A. (1966) 'Latency Relaxation: A Brief Analytical Review', *MCV Quarterly* **2**, 82–89.

Sandow, A. (1970) 'Skeletal Muscle', *Ann. Rev. Physiol.* **32**, 87–138.

Schatzmann, H. J. (1968) 'Action of Acetylcholine and Epinephrine on Intestinal Smooth Muscle', in *Handbook of Physiology*, Section 6, Vol. 4, Ed. C. F. Code, American Physiological Society, Washington, D.C.

Schofield, G. C. (1968) 'Anatomy of Muscular and Neural Tissues in the Alimentary Canal', in *Handbook of Physiology*, Section 6, Vol. 4, Ed. C. F. Code, American Physiological Society, Washington, D.C.

Schoenberg, C. F. (1973) 'The Influence of Temperature on the Thick Filaments of Vertebrate Smooth Muscle', *Phil. Trans. R. Soc. Lond. B* **265**, 197–202.

Sjostrand, N. O. (1965) 'The Adrenergic Innervation of the Vas Deferens and the Accessory Male Genital Glands', *Acta Physiol. Scand.* **65**, suppl. 257, 1–82.

Small, J. V. and Squire, J. M. (1972) 'Structural Basis of Contraction in Vertebrate Smooth Muscles', *J. Mol. Biol.* **67**, 117–149.

Smith, U. (1970) 'The Origin of Small Vesicles in Neurosecretory Axons', *Tissue and Cell* **2**, 427–433.

Sobieszek, A. (1973) 'The Fine Structure of the Contractile Apparatus of the Anterior Byssus Retractor Muscle of *Mytilus edulis*', *J. Ultrastruct. Res.* **43**, 313–343.

Sobieszek, A. and Small, J. V. (1973) 'The Assembly of Ribbon Shaped Structures in Low Ionic Strength Extracts obtained from Vertebrate Smooth Muscle', *Phil. Trans. R. Soc. Lond. B* **265**, 203–212.

Somlyo, A. P. (1972) 'Excitation-contraction Coupling in Vertebrate Smooth Muscle: Correlation of Ultrastructure with Function', *The Physiologist* **15**, 338–348.

Somlyo, A. P., Devine, C. E., Somlyo, A. V. and North, S. R. (1971) 'Sarcoplasmic Reticulum and the Temperature-dependent Contraction of Smooth Muscle in Calcium-free Solutions', *J. Cell Biol.* **51**, 722–741.

Somlyo, A. P., Devine, C. E., Somlyo, A. V. and Rice, R. V. (1973) 'Filament Organization in Vertebrate Smooth Muscle', *Phil. Trans. R. Soc. Lond. B* **265**, 223–229.

Squire, J. M. (1971) 'General Model for the Structure of All Myosin-containing Filaments', *Nature* **233**, 457–462.

Steedman, W. M. (1966) 'Microelectrode Studies on Mammalian Vascular Muscle', *J. Physiol. (Lond.)* **186**, 382–400.

Stossel, W. and Zebe, E. (1968) 'Zur intracellularen Regulation der Kontractions-aktivitat', *Pflugers Arch.* **302**, 38–56.

Syson, A. J. (1974) 'Studies on the Excitation-contraction Mechanisms of Mammalian Smooth Muscle', Ph.D. Thesis, University of Lancaster.

Syson, A. J. and Huddart, H. (1973) 'Contracture Tension in Rat Vas Deferens and Ileal Smooth Muscle and its Modification by External Calcium and the Tonicity of the Medium', *Comp. Biochem. Physiol.* **45A**, 345–362.

Szent-Gyorgyi, A. G., Cohen, C. and Kendrick-Jones, J. (1971) 'Paramyosin and the Filaments of Molluscan 'Catch' Muscle. II Native Filaments: Isolation and Characterization', *J. Mol. Biol.* **56**, 239–258.

Tomita, T. (1966) 'Membrane Capacity and Resistance of Mammalian Smooth Muscle', *J. Theoret. Biol.* **12**, 216–227.

Tomita, T. (1970) 'Electrical Properties of Mammalian Smooth Muscle', in *Smooth Muscle*, Ed. Bulbring, E., Brading, A. F., Jones, A. W. and Tomita, T., Edward Arnold, London.

Toner, P. G. and Carr, K. E. (1971) *Cell Structure*, 2nd. edition, Churchill-Livingstone, Edinburgh.

Van Der Kloot, W. G. (1965) 'The Uptake of Calcium and Strontium by Fractions from Lobster Muscle', *Comp. Biochem. Physiol.* **15**, 547–565.

Van Der Kloot, W. G. (1968) 'The Effect of Disruption of the T Tubules on Calcium Efflux from Frog Skeletal Muscle', *Comp. Biochem. Physiol.* **26**, 377–379.

Washizu, Y. (1966) 'Grouped Discharges in Ureter Muscle', *Comp. Biochem. Physiol.* **19**, 713–728.

Washizu, Y. (1967) 'The Electrical Properties of Leech Dorsal Muscle', *Comp. Biochem. Physiol.* **20**, 641–646.

Washizu, Y. (1968) 'Conversion of Ureteral Action Potentials by Metal Ions', *Comp. Biochem. Physiol.* **24**, 301–305.

Washizu, Y. (1969) 'Potassium and Sodium Content of Guinea-pig Ureteral Muscle', *Comp. Biochem. Physiol.* **28**, 425–429.

Weber, A. (1968) 'The Mechanism of Action of Caffeine on Sarcoplasmic Reticulum', *J. gen. Physiol.* **52**, 760–772.

Weber, A. and Herz, R. (1968) 'The Relationship between Caffeine Contracture of Intact Muscle and the Effect of Caffeine on Reticulum', *J. gen. Physiol.* **52**, 750–759.

Wyburn, G. M. (1960) *The Nervous System*, Academic Press, London.

Subject Index